Advances in the Computational Sciences

Symposium in Honor of

Dr Berni Alder's 90th Birthday

Advances in the Computational Sciences

Symposium in Honor of **Dr Berni Alder's 90th Birthday**

Lawrence Livermore National Laboratory, 20 August 2015

Editors

Eric Schwegler

Brenda M. Rubenstein

Stephen B. Libby

Lawrence Livermore National Laboratory, USA

EW JERSEY • LONDON • SINGAPORE • BEIJING • SHANGHAI • HONG KONG • TAIPEI • CHENNAI • TOKYO

Published by

World Scientific Publishing Co. Pte. Ltd.
5 Toh Tuck Link, Singapore 596224
USA office: 27 Warren Street, Suite 401-402, Hackensack, NJ 07601
UK office: 57 Shelton Street, Covent Garden, London WC2H 9HE

British Library Cataloguing-in-Publication Data
A catalogue record for this book is available from the British Library.

ADVANCES IN THE COMPUTATIONAL SCIENCES
Proceedings of the Symposium in Honor of Dr Berni Alder's 90th Birthday

ISBN 978-981-3209-41-1

Desk Editor: Christopher Teo

Printed in Singapore

Contents

PREFACE

On August 20, 2015, a symposium at LLNL was held in honor of Berni J. Alder's 90th birthday. Many of Berni's scientific colleagues and collaborators, former students, and post-doctoral fellows came to celebrate and honor Berni and the ground-breaking scientific impact of his many discoveries.

Since the late 1940's, when Berni studied at Caltech under John G. Kirkwood, he has been at the forefront of the development and application of computational molecular dynamics and Monte Carlo methods for the study of the microscopic properties of matter. In fact, Berni and his collaborators have the rare distinction of having truly discovered new, and quite unforeseen, physical phenomena in numerical simulation. Two well-known examples, both done in collaboration with Tom Wainwright, were the discoveries of a melting transition in the 2D hard disk system, and the discovery of power laws – long time tails, in fluid relaxation. The former directly contradicted the Mermin-Wagner theorem (and arguments due to Landau) that there could be no phase transitions in 2D systems with continuous symmetry, while the latter upended the Boltzmann idea that, due to uncorrelated scattering, non-equilibrium fluid systems would relax to equilibrium exponentially quickly. These discoveries were not mere isolated exceptions to the established orthodoxy. Rather, they were the beginning of wonderful lines of inquiry that ultimately greatly enlarged our view of phase transitions, and the true microscopic dynamics of non-equilibrium phenomena. Similarly, Berni's and David Ceperley's 1980 quantum Monte Carlo study of the correlation energies of the electron gas well beyond the Gell-Mann and Brueckner high-density regime was (and is still) an important enabler of accurate applications of density functional theory to real materials.

Many of the topics chosen by the symposium speakers and contributors to this volume give a good sense of their current, cutting edge work, often growing out of these discoveries, illuminating the microscopic statistical behaviors of matter.

Daan Frenkel's talk, for example, connected early work on non-equilibrium processes (always keeping an eye on the role of entropy) with his current work on nano-assembly of complex structures –all emerging from simple rules. Jan Sengers then discussed long-range correlations in fluids, mode coupling, and the importance of fluctuations. All of these topics were themes that would recur throughout the symposium. Bill Hoover's talk interwove his own scientific trajectory at the lab with Berni's scientific influence – first in their hard disk phase transition studies, and on to many innovations in physically motivated algorithms for the numerical analysis of general non-equilibrium phenomena, classically chaotic systems, and shock waves (in particular).

J. Robert Dorfman's talk then beautifully detailed the development of the new ideas on non-equilibrium fluid relaxation that began with theoretical insights in the 1960's. Spurred by the high density simulation results of the 1970 Alder-Wainwright paper, these ideas led to a correct kinetic explanation of the dimension dependent 'long time tails' scaling like ~ t-d/2.

Another major theme of the symposium (and, indeed, of Berni's current research) was the role of fluctuations in hydrodynamics and diffusion. Dov Shvarts's covered current leading work, a significant portion from his own group, on the analysis of turbulent flow initiated by interface mixing instabilities encountered, for example, in inertial fusion. Then, Todd Weisgraber discussed molecular dynamics simulations of the transition to turbulence and the emerging significance of smooth versus rough boundary conditions. In his talk, Aleks Donev showed how a more careful treatment of fluctuations leads to the prediction of giant mesoscopic fluctuations in fluids whose ensemble averages still obey Fickian diffusion and the Einstein-Stokes relations. And, as explained by Tim Germann in his talk, with the advent of exascale computing, this area is likely ripe for discovery – with careful numerical simulation leading the way.

Algorithmic developments in atomistic and quantum mechanical simulations were another major theme of our conference, with new developments in molecular dynamics being described by Alex Garcia, and a new approach to the 'fermion sign problem' in quantum Monte Carlo presented by Jonathan DuBois. In the latter arena, David Ceperley covered Berni's important role in the development of quantum Monte Carlo, from their key paper in 1980 to the present.

Charles Bennett, Berni's former student from the late 1960's, and a major innovator in the field of quantum information, gave a fascinating, thought provoking talk about the interplay of computational complexity and human ethics.

Robert Laughlin, a colleague of Berni's from his time at LLNL in the 1980's and a Nobel laureate, stretched our imaginations in a different direction, presenting his latest thinking about the explanation of form and function in biology with his key idea being that biological systems represent successful 'engineered' solutions with the tools at hand – such as the possible use of reaction-diffusion systems at criticality to function as a needed stable length measurement, for example.

Finally, as our current and former directors Bill Goldstein and Bruce Tarter stressed in their remarks, Berni not only had dramatic worldwide scientific impact, but also a significant local impact on scientific computing, the equation of state program at LLNL, and graduate education at the University of California Davis Department of Applied Science.

This volume also contains several fascinating contributions from several of Berni's colleagues, some of who weren't able to attend the symposium. These include Masaharu Isobe's overview of recent advances in simulating hard sphere systems with event driven molecular dynamics techniques, Jean-Pierre Hansen's discussion of the current state of understanding of glass transition dynamics, Mike Kosterlitz's personal history of his and David Thouless's discovery of phase transitions in 2D systems with continuous symmetry, which was wonderfully honored by this year's Nobel Prize in Physics, and the role Berni's simulations played, Sidney Yip's paper on the state of understanding of the behavior of viscosities in supercooled liquids, Raymond Jeanloz' recent work on broadly applicable finite strain relations and the shock Hugoniots of many materials, George Chapline's thought- provoking ideas about possible uses of quantum mechanics in pattern recognition, and Mary Ann Mansigh Karlsen's charming reminiscences of her 25 years working with Berni on molecular dynamics.

Stephen B. Libby
Physics Division
Lawrence Livermore National Laboratory
Livermore, California

Brenda M. Rubenstein
Department of Chemistry
Brown University
Providence, Rhode Island

Eric Schwegler
Materials Science Division
Lawrence Livermore National Laboratory
Livermore, California

8:30-8:45 a.m.	Welcoming	Dr. Bruce Tarter
Session 1, Chair: Dr. Bruce Tarter		
8:45-9:15	Simple Models, Complex Structures	Prof. Daan Frenkel University of Cambridge
9:15-9:45	Long-range Correlations in Fluids	Prof. Jan Sengers University of Maryland
9:45-10:15	An appreciation of Berni Julian Alder	Dr. Bill Hoover Ruby Valley Research Institute
10:15-10:30	*Coffee Break*	
Session 2, Chair: Prof. Richard Scalettar		
10:30-11:00	Computation, Ethics, and Cosmology	Dr. Charlie Bennett IBM Thomas J. Watson Research Center
11:00-11:30	Non-Equilibrium is Different	Prof. J. Robert Dorfman University of Maryland
11:30-12:00	Quantum Monte Carlo and Berni Alder	Prof. David Ceperley University of Illinois at Urbana Champaign
12:00-1:30 p.m.	*Lunch*	
Session 3, Chair: Prof. Phillip Colella		
1:30-2:00	The Poor Man's Molecular Dynamics	Prof. Alex Garcia San Jose State University
2:00-2:30	Nanohydrodynamics and Exascale Frontiers in Molecular Dynamics Simulation	Dr. Tim Germann Los Alamos National Laboratory
2:30-3:00	Late Time Evolution of Hydrodynamic Instabilities and Turbulent Mixing	Dr. Dov Shvarts Nuclear Research Center Negev
3:00-3:30	The Journey to Biology	Prof. Robert Laughlin Stanford University
3:30-3:45	*Break*	
Session 4, Chair: Prof. Robert Laughlin		
3:45 - 4:15	Computer Experiments on the Onset of Turbulence	Dr. Todd Weisgraber Lawrence Livermore National Laboratory
4:15-4:45	The truth about diffusion (in liquids)	Prof. Aleks Donev Courant Institute of Mathematical Sciences
4:45-5:15	Strategies for overcoming the fermion sign problem	Dr. Jonathan DuBois Lawrence Livermore National Laboratory
5:15-5:30	Closing Remarks	Dr. William Goldstein
5:30-6:30	*Reception*	

Lawrence Livermore National Laboratory
7000 East Avenue · Livermore, CA 94550 · (925) 422-1100 · http://www.llnl.gov
Operated by Lawrence Livermore National Security, LLC for the Department of Energy's National Nuclear Security Administration

Chapter 1

Welcome and Reflections on Berni Alder

C. Bruce Tarter

Director Emeritus
Lawrence Livermore National Laboratory

Good morning and welcome to Berni Alder's 90th Birthday Symposium. My name is Bruce Tarter and I'm representing Director Bill Goldstein, who could not be here this morning, but will be here for the afternoon session and will make comments of his own at the end of the Symposium.

I will note that this is the third time in recent years we have honored a Lab person who is still contributing well into their 90s. Three years ago, we celebrated former Lab Director Johnny Foster's 90th birthday, and more recently we recognized Dick Post in his 90s with a lifetime achievement award for his continuing professional efforts in fusion science and technology. Today, we have the great pleasure of honoring Berni Alder's extraordinary career on the occasion of his 90th birthday. I draw one simple conclusion from these events: as a Lab scientist you can expect a long life – but only if you continue to work productively.

Now, I will reflect briefly on my own association with Berni. I came to the Lab in 1967 and joined Theoretical ("T") Division where Berni organizationally resided. Berni was already outside the 'normal' structure of the Division, and I was somewhat in awe of him as a member of the National Academy of Science (the only Lab person, except for Edward Teller, with that distinction). I didn't quite understand what Berni and his research group did, but Dave Young, a recent post-doc in his group explained the science of molecular dynamics (MD) to me over lunches that first year as well as how the Lab computers were the special enabling tool for the work (I had also come to the Lab in part because I had been trained as a computational astrophysicist).

A few years later, in 1978, I became Head of T-Division and was now the custodian of Berni's special relationship with the computers and the computational system at the Lab. Many years earlier, Berni and Sid Fernbach, the long term Head of Computation, had invented a remarkable approach to running the small memory, but long running MD calculations crucial to the research. Specifically, if done properly, the MD code could fit into a small part of the computer and start and stop automatically. Consequently, it could exploit both open areas in memory as well as short periods when nothing else was running. I should add here that it was only due to the superlative efforts of Mary Ann Mansigh and other members of Berni's group that this was achieved.

Since this did not negatively impact anything else on the computers, Sid essentially gave Berni the computer time for free. By the time I was the responsible administrator, there was a nominal small charge (since the weapons program could never quite believe it was truly a free lunch), but nonetheless MD research was able to garner huge amounts of computer time at an incredibly small cost (that would have run into the unaffordable millions if charged at the published rates). The other by-product was that it demonstrated even greater efficiency and use of the computers so that Sid and the Lab were able to argue to the DOE that we needed newer and larger computers!

So, as you all know, this resulted in many years of state-of-the art research with the MD tools that Berni and his group developed and applied and we will hear about this throughout today's symposium. But, before beginning that program, I want to note one other major area in which Berni personally had a great impact. As you all know, we still hear the silly phrase, "there is now emerging a third branch of physics, computational, that will join theoretical and experimental as a fundamental pillar of physics." Of course, this began to happen nearly 50 years ago and became true in practice several decades ago, and Berni played a significant role in making that happen. In the 1960s and 70s, he and Sid Fernbach pioneered the series of books, "Methods in Computational Physics" that became the standard reference for virtually everyone engaged in numerical computation. At the same time, they started the "Journal of Computational Physics" to provide a place for original research, and the editorial offices were run at Livermore for many years. Finally, both Berni and Sid were instrumental in helping create the Division of Computational Physics within the American Physical Society. So, in my mind, Berni's legacy is not just his MD research, but also his special place in the broad history of computational science. It is a great pleasure to have known Berni for nearly 50 years and to begin today's program in honor of his 90th birthday.

Chapter 2

An Appreciation: Berni Julian Alder

William Graham Hoover

Ruby Valley Research Institute
Highway Contract 60, Box 601
Ruby Valley, NV 89833 USA

Abstract

Berni Alder profoundly influenced my research career at Lawrence Livermore National Laboratory and the Davis Campus' Teller Tech, beginning in 1962 and lasting for over fifty years. I very much appreciate the opportunity provided by his Ninetieth Birthday Celebration to review some of the many high spots along the way.

2.1 Ann Arbor Revisited and Durham Explored

My Father, Edgar Malone Hoover, Junior, taught economics at Harvard (PhD, 1928) and the University of Michigan until World War II brought him to Washington for work with the National Resources Planning Board, the Office of Price Administration, and the Office of Strategic Services. After half a dozen of those years in Washington, followed by a Chemistry major at Oberlin College (AB, 1958), I returned to Ann Arbor for a PhD in Chemical Physics, 1958-1961. Three of the scientific highlights of those years were [1] a short course in FORTRAN (a three-hour lecture, taught in a single evening); [2] George Uhlenbeck's lectures on *Gastheorie*, delivered

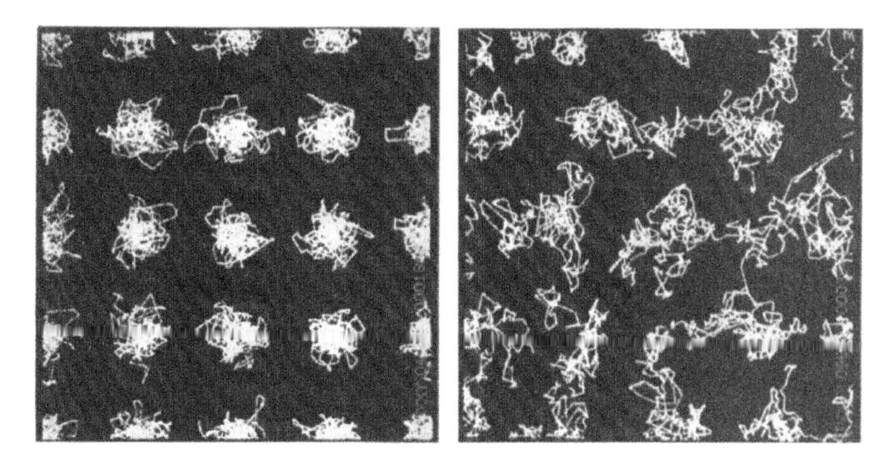

Fig. 2.1 Particle trajectories of 32 hard spheres with spatially periodic boundary conditions. At densities less than two-thirds of close packing, only the fluid phase is stable. From reference [Alder and Wainwright (1959)].

while holding his musty notes at arm's length; and [3] Andrew De Rocco's course on statistical mechanics. I was specially inspired by the computer-generated pictures in Berni and Tom Wainwright's "Molecular Motions," published in the 1959 Scientific American [Alder and Wainwright (1959)]. See Fig. 2.1. When I saw their work, I wanted to make some of these many body dynamics pictures myself.

In the early 1960s, the main theoretical route to equations of state was through integral equations for the pair distribution function. This elegant and absorbing approach was soon made thoroughly obsolete by the two simulation techniques of molecular dynamics and Monte Carlo. In 1961-1962, I spent a year performing postdoctoral work at Duke with one of John Kirkwood's students, Jacques Poirier. The result was an improved understanding of integral equations and the Mayers' virial series. At Duke, I shared an office with Jacques' student, John Nelson Shaw, whose Ph D project was the development of a one-component-plasma Monte Carlo code.

Each of the variables in John's computer program was named for a member of his Family. The code was used soon after by Steve Brush, Harry Sahlin, and Edward Teller at the Livermore Laboratory. Computing was slow in those days. Moore's 1965 Law was not yet known. Automatic equation-of-state calculations, though soon to be commonplace, were still a few years away in the unforeseeable future.

2.2 Los Alamos and Livermore

When it came time to find a "real job," I was still motivated by the Alder-Wainwright Scientific American article and applied to both Livermore and Los Alamos, where the best computers were. I had interview talks with Bill Wood at Los Alamos and Berni at Livermore. Both places were appealing with rather different physical environments but wonderful opportunities for computational research. The higher salary offer (though with much shorter vacations) brought me to Livermore, providing the chance to do real simulations rather than follow the not-so-reliable and not-so-simple integral-equation and virial-series paths.

My first California publication, with Berni and Tom, [Alder *et al.* (1963)] showed that two hard disks, with periodic boundary conditions, give a van der Waals' pressure-volume loop within a few percent of the large-system transition that had been Berni's interest ever since his doctoral work with Kirkwood. See Fig. 2.2. Research at Livermore in the early 1960s was a joy, free from the need to apply for grants or to write progress reports. My efforts were strengthened by stimulating collaborations with Francis Ree, Tom Wainwright, and Berni. Often our relatively-long research days were divided up by dinner discussions at Livermore's Yin Yin Restaurant (1960 - present). In those days, abalone was still on the menu.

2.3 Work with Francis Ree

Francis Ree was a student of Henry Eyring's at the University of Utah, mathematically gifted, and a perfect coworker. He and I had the idea to implement Kirkwood's "communal-entropy" model into Monte Carlo simulations measuring hard-disk and hard-sphere entropies. Kirkwood's idea was that fluid phases enjoyed an additional shared entropy Nk as a consequence of indistinguishability. This shared entropy was identified with an extra e^N in fluid partition functions. With Edward Teller's permission to use the considerable computer time entailed, Francis and I measured the

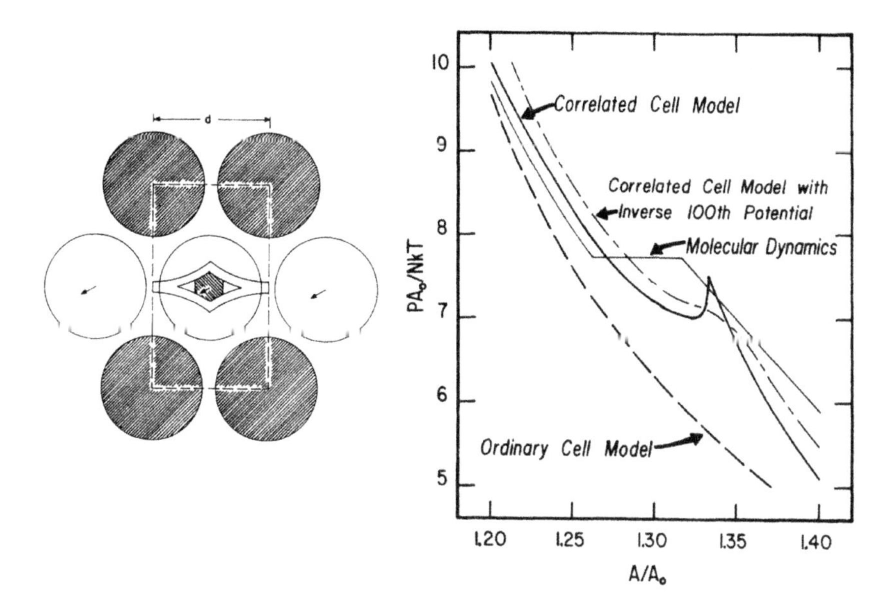

Fig. 2.2 On the left, a row of hard disks moves cooperatively, expanding the "free volume" explored by about one third and allowing a phase transition as that periodic row slips by its neighbors. Without this cooperative motion, the "ordinary" cell model predicts a pressure somewhat lower than the results of the many body simulations shown at the right. From reference [Alder *et al.* (1963)].

communal entropy for both disks and spheres. We found that the "extra" fluid entropy varies slowly with density rather than appearing suddenly at melting. We used quantitative measurements to compute precisely the density dependence of the entropy difference between the fluid and solid phases. Our communal-entropy work, [Hoover and Ree (1968)] along with solid-phase investigations carried out with Berni and David Young, [Alder *et al.* (1968)] led to accurate locations of the melting transitions for both disks and spheres.

By the late 1960s, Francis and I had had enough virial series and entropy work on the hard-disk and hard-sphere problems. We sought out new directions. Francis enjoyed phase diagram work while I, now working for Russ Duff, pursued nonequilibrium studies. In 1967 I was the very last one of seven authors on a shockwave paper presented by Russ in Paris [Duff *et al.* (1968)]. This work documented progress toward simulating shock-waves with *continuous-potential* molecular dynamics. The feasibility of such simulations had been established by Enrico Fermi (at Los Alamos), George Vineyard (at Brookhaven), as well as Aneesur Rahman (at Argonne). I had missed the Fermi and Vineyard work. But, Rahman's later work did get my attention [Rahman (1964)] and I implemented his predictor-corrector algorithm.

Brad Holian, Bill Moran, Galen Straub, and I [Holian *et al.* (1980)] revisited shockwave simulations in the 1980s, using the much more efficient leapfrog algorithm. That work focused on the strong compression of liquid states, as is illustrated in Fig. 2.3. Decades later, by then converted to Runge-Kutta, and working with Paco Uribe and my Wife Carol, [Hoover *et al.* (2010)] we studied more detailed models of the shock process. These models used a tensor temperature, with $T_{xx} >> T_{yy} = T_{zz}$. We also found and included time delays between the fluxes of momentum and energy and the velocity and temperature gradients driving these fluxes. Fig. 2.4 shows the pressure tensor and the heat-flux vector for a simple model including these effects. Such a model can describe numerical molecular dynamics data rather well [Hoover *et al.* (2010)].

2.4 Bill Ashurst and Nonequilibrium Molecular Dynamics

Once several workers (Hans Andersen, John Barker, Frank Canfield, David Chandler, Doug Henderson, Ali Mansoori, Jay Rasaiah, George Stell, and John Weeks) had developed a hard-sphere-based perturbation theory of equilibrium liquid states, *nonequilibrium* simulations opened up as a more promising source of new ideas. In 1971, Berni had helped me into a part-time Professorship at U C Davis through Edward Teller's Department of Applied Science. My first Ph D student there, Bill Ashurst, from Sandia's Livermore Laboratory, became interested in modeling shear and heat flows directly, with *thermostated* "nonequilibrium molecular dynamics" [Hoover and Ashurst (1975)].

Bill used differential feedback to maintain constant boundary velocities and temperatures. Some of his transport coefficients disagreed with those computed from Green-Kubo linear-response theory (but with factor-of-two errors) by our French colleagues, Levesque, Verlet, and Kürkijarvi [Levesque *et al.* (1973)]. Thanks to a grant from the Academy of Applied Science (Concord, New Hampshire), made possible through my U C Davis position,

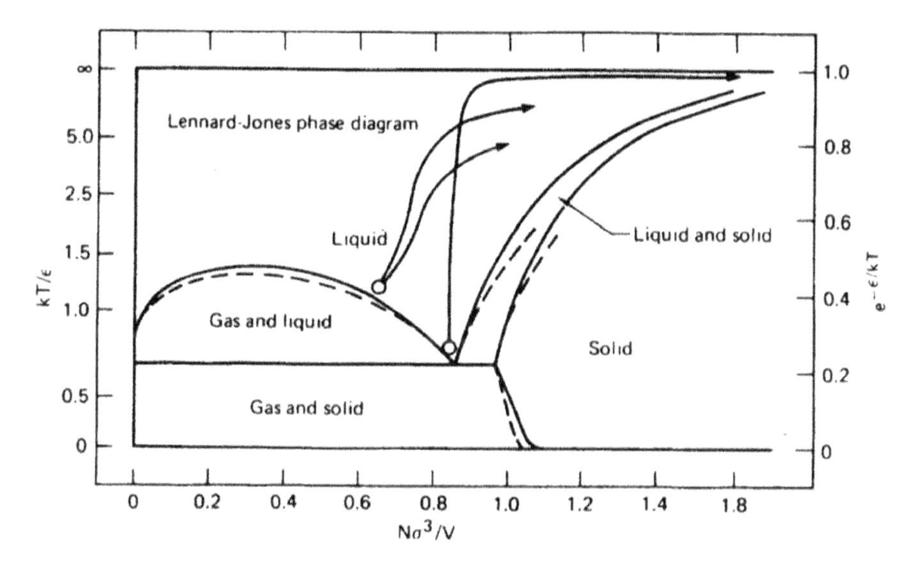

Fig. 2.3 The temperature-density trajectories of three shockwave simulations starting in the liquid phase and ending up as highly-compressed hot gas. The dashed lines represent the phase diagram of liquid argon. From reference [Holian *et al.* (1980)].

I was able to finance summer research projects with bright high school students. The same grant enabled considerable foreign travel so that I could exchange ideas with the many researchers gathered together in France. Our interests in common with these researchers led to numerous trips to Paris and Orsay to participate in Carl Moser's simulation workshops. The European contacts were stimulating and pleasant and led to a productive sabbatical in Wien and a thirty-year collaboration with Harald Posch and his colleagues.

Some of the puzzles that arose from this 70s-80s-era work remain today. For example, the effective viscosity measured in a strong shockwave exceeds the small strain rate Newtonian one by tens of percent. At the same time, a steady *homogeneous* shear flow at the same strain rate exhibits a *reduced* viscosity, also by tens of percent. These opposite nonlinear effects indicate that there is still much to learn about nonlinear transport.

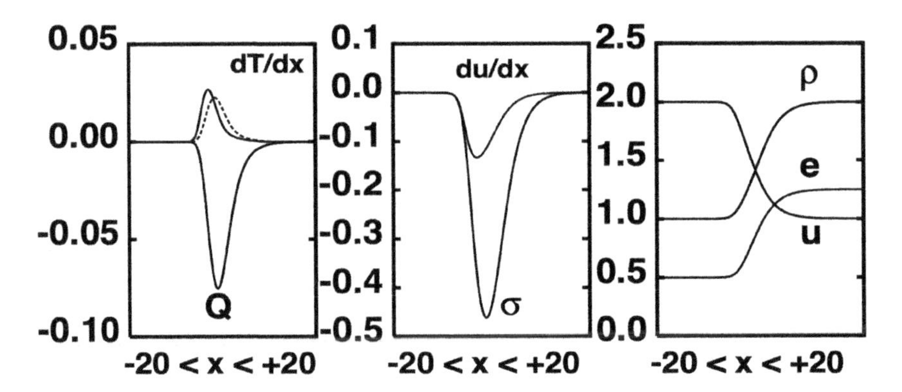

Fig. 2.4 The time delay between stress σ and strain rate (du/dx) and between heat flux Q and the temperature gradients (dT/dx) (longitudinal and transverse) are shown for continuum simulations of moderately-dense shockwaves. These models describe the results of molecular dynamics simulations quite well. The figure is from reference [Hoover *et al.* (2010)].

2.5 Sabbatical Research in Australia, Austria, and Japan

My academic connection to the University of California at Davis made it possible to have sabbaticals — Canberra (1977-1978), Wien (1985), and Yokohama (1989-1990). The year in Australia enabled my son Nathan, just graduated from Livermore High, to work with me, exploring another of Berni's interests, hard-disk and sphere "free volumes." Kenton Hanson did the three-dimensional work in Berkeley. Nathan and I discovered a percolation transition for hard disks, where the free volume changes from extensive to intensive, at one-fourth of the hard-disk close-packing density [Hoover *et al.* (1979)]. See Fig. 2.5. It was interesting to see that solid-phase free volumes are larger than the fluid ones at the same density.

In Wien, I worked with Karl Kratky, Harald Posch, and Franz Vesely while lecturing and writing my first book, "Molecular Dynamics." The year in Japan, arranged by Shuichi Nosé, was shared with my new wife, Carol, resulting in a longer sequel book, "Computational Statistical Mechanics" with inspiration and cover art from Shuichi (Yokohama) and Harald Posch (Wien). See Fig. 2.6. Nosé's work changed my outlook on computer simulation. Let us look at some of the details.

2.6 Shuichi Nosé and Keio University

During Orwell's "1984," I came across two amazing papers [Nosé (1984a,b)] in the Library of the Livermore Laboratory. They were written by a then-unknown Japanese postdoc, Shuichi Nosé, who was working in Canada with Mike Klein. The title of one of Shuichi's papers [Nosé (1984a)] describes the gist of his work, "A Molecular Dynamics Method for Simulations in the Canonical Ensemble." In 1984, this title's concept seemed to me completely paradoxical. For me, "Molecular Dynamics" had generally meant the *micro*canonical constant-energy ensemble, not the very different constant-temperature canonical one. Although Bill Ashurst and I had long carried out "isokinetic" simulations for over a decade, the notion of a "canonical" dynamics made no sense to me. I set out to meet Nosé at an upcoming workshop in France and was lucky enough to find him, completely by accident, in Paris' Orly train station days prior to the workshop's start.

Nosé's novel thermostat ideas took me a couple of weeks to digest back then in 1984, even after several hours of conversation before and during the CECAM (European Center for Atomic and Molecular Calculations) workshop [Hoover (1985)]. Following up Nosé's new ideas, applied to the harmonic oscillator, revealed a troubling aspect of his new dynamics. The results for long-time averages depended strongly on the initial conditions. A harmonic oscillator, in one dimension and with unit force constant and mass, was the simplest illustration. I could see that his four motion equations, which included a completely novel "time-scaling" variable s, its conjugate momentum ζ, and a "number of degrees of freedom" $\# = 2$, one for (qp) and one for $(s\zeta)$:

$$\{ \dot{q} = (p/s^2) \; ; \; \dot{p} = -q \; ; \; \dot{s} = \zeta \; ; \; \dot{\zeta} = (p^2/s^3) - (\#/s) \} \; [\text{ N }] \; , \qquad (2.1)$$

could be replaced by an equivalent but simpler and less stiff set of just *three* equations, with s absent and $\# = 1$:

$$\{ \dot{q} = p \; ; \; \dot{p} = -q - \zeta p \; ; \; \dot{\zeta} = p^2 - \# \} \; [\text{ NH }] \; . \qquad (2.2)$$

Fig. 2.5 The top row shows exclusion disks which the centers of other disks cannot penetrate. The lower row shows the particles themselves and their free volumes at a density near that of the fluid-solid transition. The pictures in the top row make it plausible that there is a percolation transition (from extensive free volumes to intensive) between the densities shown. In reference [Hoover *et al.* (1979)], numerical work shows that the transition density is close to one-fourth of the close-packed density.

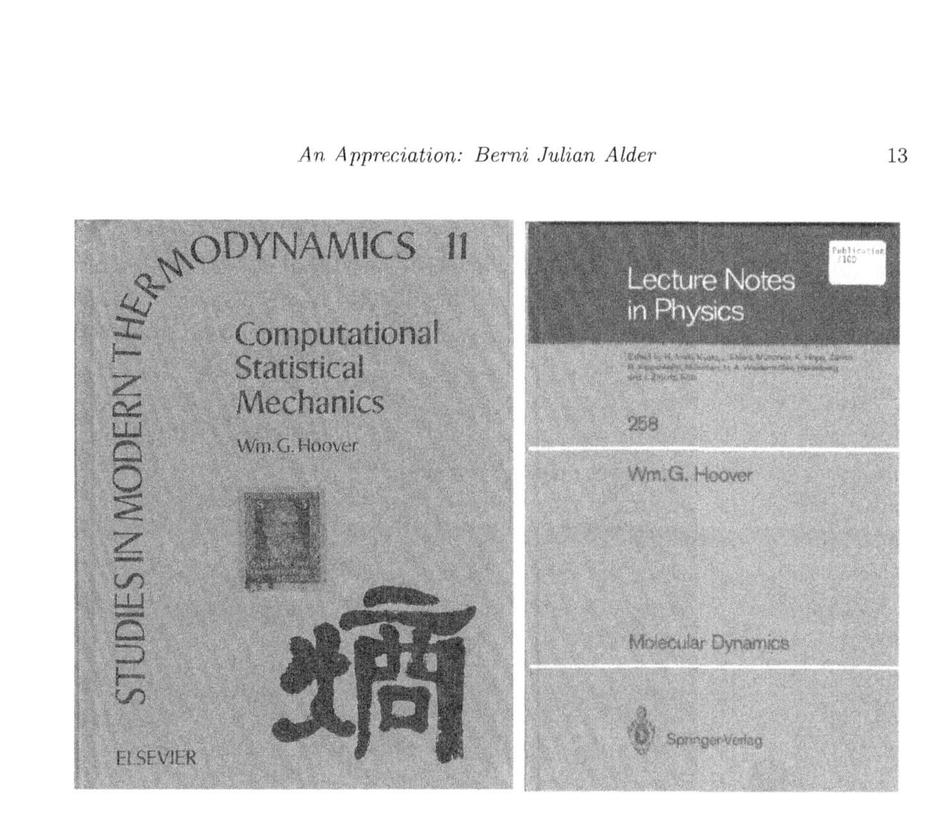

Fig. 2.6 Two books written during sabbaticals in Austria and Japan. The two-part Japanese character combining "heat" and "divide" represents "entropy." Note Boltzmann's presence here.

Solutions of these three "Nosé-Hoover" equations give $\zeta(q)$ trajectories *identical* to those from the four "Nosé" equations provided that (1) the #s match, (2) the initial value of s is unity, and (3) the other initial values match: $(qp\zeta)_{\rm N} = (qp\zeta)_{\rm NH}$. And it was true, as Nosé had pointed out in his papers, that the three equations were likewise consistent with Gibbs' canonical distribution, along with an additional Gaussian distribution for the momentum variable ζ :

$$f(q,p,\zeta) \propto e^{-q^2/2}e^{-p^2/2}e^{-\zeta^2/2} \ . \tag{2.3}$$

In fact, by starting with Gibbs' distribution and working backward using Liouville's phase-space flow equation, I could obtain Nosé's "time-scaled" dynamics without any consideration of time scaling at all!

Besides the mysterious time-scaling, there was still that troubling fly in the ointment. The thermostated harmonic-oscillator dynamics doesn't actually generate *all* of Gibbs' distribution. Instead six percent of its Gaussian measure occupies a "chaotic sea" in which nearby trajectories separate from one another with a positive exponential growth rate proportional to $e^{+\lambda t}$ where $\lambda = 0.0139$ is the system's largest "Lyapunov exponent" [Hoover *et al.* (2015a)]. All of the remaining 94 percent of the Gaussian distribution is occupied by an infinite set of periodic orbits and their tori. Each such periodic orbit is surrounded by concentric stable toroidal orbits with *vanishing* Lyapunov exponents, $\lambda = 0$. The complexity of this distribution can be visualized in the $(0p\zeta)$, $(q0\zeta)$, and $(qp0)$ cross sections of Fig. 2.7. Understanding all of this new information evolved gradually, over a period of years rather than days.

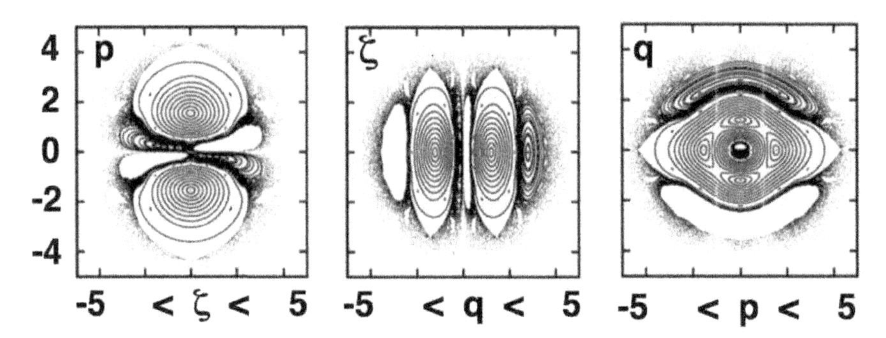

Fig. 2.7 Cross sections of the chaotic sea with a selection of toroidal closed-curve orbits for the equilibrium Nosé-Hoover oscillator with $T = 1$. From reference [Hoover *et al.* (2015a)].

Ultimately, meeting Shuichi Nosé in Paris in 1984 led me into many new research directions running the gamut from one-body chaos to many-body hydrodynamics, combining ideas from dynamical systems and chaos theory with continuum mechanics and molecular dynamics. In turn, this led to a strong and prolific collaboration (50 joint papers) with Harald Posch at Boltzmann's University on Boltzmanngasse in Wien. By this time, I had moved away from Berni's research interests, which had become mainly quantum mechanical. My own work with Harald, and later, with Carol, involved thermostats and the connections linking continuum mechanics to molecular dynamics and dynamical systems theory.

Many of these projects involved Brad Holian, a gifted student of Berni's and a real enthusiast for *many*-body molecular dynamics. At the other extreme, a single degree of freedom, Brad and I were able to show that the troubling one-dimensional harmonic oscillator, with not only its second velocity moment controlled, but also its fourth, *is* "ergodic" [Hoover and Holian (1996)]. That is, with both velocity moments controlled time reversibly, this four-equation model provides *all* of Gibbs' canonical distribution for the oscillator as well as Gaussian distributions for the two thermostat variables ζ and ξ :

$$\{ \dot{q} = p \; ; \; \dot{p} = -q - \zeta p - \xi(p^3/T) \; ; \; \dot{\zeta} = (p^2/T) - 1 \; ; \; \dot{\xi} = (p^4/T^2) - 3(p^2/T) \}$$
$$\longrightarrow (2\pi)^2 T f(q,p,\zeta,\xi) \equiv e^{-q^2/2T} e^{-p^2/2T} e^{-\zeta^2/2} e^{-\xi^2/2} \; . \qquad (2.4)$$

The idea of formulating a canonical-ensemble dynamics (giving a Gaussian velocity distribution defined by its kinetic temperature $\langle \; p^2 \; \rangle$ rather than constant energy) was an outgrowth of Shuichi Nosé's 1984 Magicianship [Nosé (1984a,b)]. Carol and I had married in 1988 in preparation for a very productive year (1989-1990) together at Nosé's Keio University. The working conditions in Japan were extremely pleasant: I walked to work at Keio University's Hiyoshi campus and had no real duties other than speaking at a few conferences during the year. We collaborated with Tony De Groot who had built a CRAY-speed computer back at Livermore with a transputer budget of only $30,000. With many colleagues' help we were able to simulate plastic flow with millions of degrees of freedom in reasonable clock times with Tony's machine. See Fig. 2.7 [Hoover *et al.* (1990)].

2.7 Recent Studies of Ergodicity and Lyapunov Instability

During 2014-2015, Carol and I had been enjoying a very fruitful collaboration with Clint Sprott (Wisconsin) and Puneet Patra (a PhD student

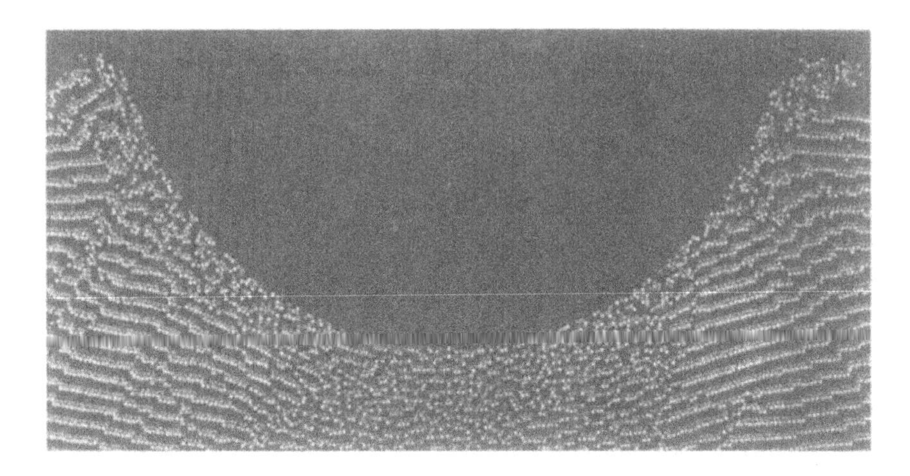

Fig. 2.8 Plastic flow around an indentor with more than a million atoms. A simulation carried out on Tony De Groot's low-cost transputer computer, the "SPRINT". From reference [Hoover *et al.* (1990)].

of Baidurya Bhattacharya's in Kharagpur) [Hoover *et al.* (2015b,c, 2016)]. We studied the ergodicity and Lyapunov instability of oscillators exposed to a temperature gradient, $T(q) = 1 + \epsilon \tanh(q)$. Over the course of a year, we came upon *several* ergodic oscillator models with only *three* motion equations, rather than four. This was a pleasant surprise. The analysis of three-dimensional topology rather than four- is a tremendous simplification.

The simplest of all these three-dimensional models incorporates "weak" (integrated) control of both p^2 and p^4 :

$$\dot{q} = p \; ; \; \dot{p} = -q - 0.05\zeta p - 0.32\zeta(p^3/T) \; ; \tag{2.5}$$

$$\dot{\zeta} = 0.05[\ (p^2/T) - 1\ ] + 0.32[\ (p^4/T^2) - 3(p^2/T)\ ] \; . \tag{2.6}$$

For *any* initial condition, the Gaussian velocity distribution results. And, now the chaotic sea embraces the entire three-dimensional phase space.

Fig. 2.9 displays the sign (red positive and green negative) of the local Lyapunov exponent just at the moment that the equilibrium $(qp\zeta)$ trajectory crosses the phase plane $\zeta = 0$. The "local" Lyapunov exponent describes the instantaneous growth or decay rate of a small displacement in the neighborhood of a "reference trajectory". In the isothermal, constant-T case, an amazing consequence of the three thermostated equations of motion is a simple three-dimensional Gaussian distribution generated by an incredibly-complicated Lyapunov-unstable, but time-reversible, one-dimensional trajectory. Fig. 2.9 conceals a surprise. If a $(qp\zeta)$ oscillator is viewed in a mirror perpendicular to the q-axis, both q and p change sign, corresponding to inversion of the cross section through the origin at its center. Diverging trajectories viewed in a mirror likewise diverge. There is also a *missing* symmetry. One would think that running a trajectory backwards, with q unchanged and p reversed in sign, would replace divergence by convergence so that red in the lower half plane would correspond to green above, and *vice versa* . This symmetry, though fully consistent with the time-reversible motion equations, is absent. The reason is that the tendency toward divergence or convergence of two trajectories, constrained by a tether, *can only depend upon the past*, and *not the future*. This relatively subtle distinction can be clarified by considering the *nonequilibrium* case where T *varies* with the coordinate q. We turn to that case next.

In the *nonequilibrium* case with temperature $T(q) = 1 + \epsilon \tanh(q)$ – so that the maximum temperature gradient is ϵ – the equations remain time-reversible. Reversing time (as well as the signs of p and ζ) could be expected to reverse the sign of the Lyapunov exponent. Fig. 2.9 shows that this natural expectation is unwarranted and wrong. The top and bottom halves of the figure are *not* simple mirror images. Evidently the time-reversiblity of all the equations is misleading. Let us consider this observed symmetry breaking in more detail.

2.8 Symmetry Breaking in Time-Reversible Flows

From the conceptual standpoint, an interesting paradoxical aspect of (irreversible) nonequilibrium flows is the time reversibility of their underlying motion equations. This contrast motivates Loschmidt's and Zermélo's Paradoxes. Loschmidt's is that forward and backward movies of flows *satisfying* the irreversible Second Law of Thermodynamics, are described by exactly the same time-reversible motion equations in both time directions.

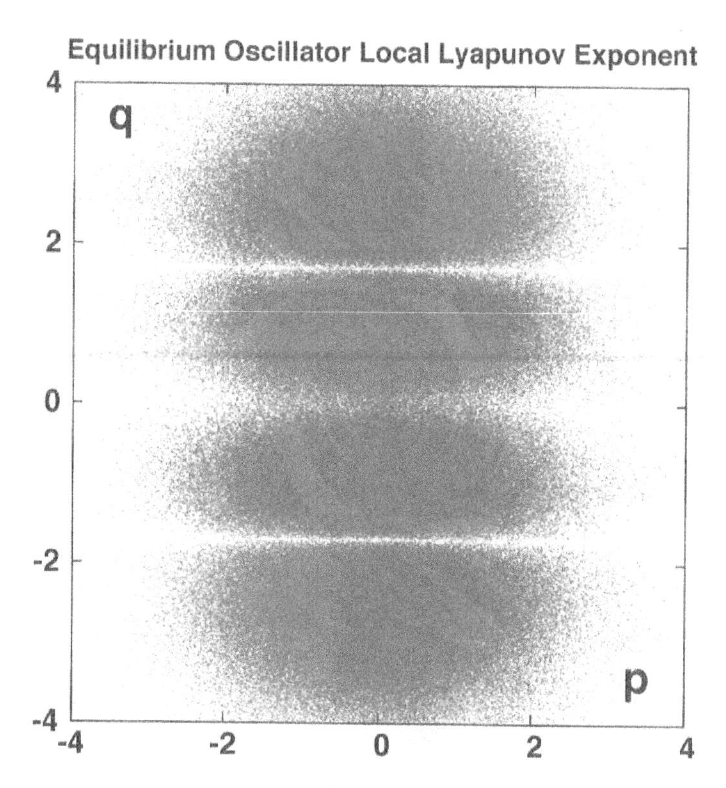

Fig. 2.9 Local Lyapunov exponents (red for positive and green for negative) for the three-equation equilibrium ergodic oscillator described in the text. The imposed constant-temperature profile is $T(q) = 1$. Notice that the color for $(\pm q, \pm p)$ is identical to that for $(\mp q, \mp p)$, corresponding to the mirror (inversion) symmetry of the problem. Time-reversal symmetry of the Lyapunov exponent is *missing* as explained in the text.

Zermélo's is that any initial phase-space state, no matter how unlikely or odd, will eventually recur due to the bounded nature of the constant-energy phase space. Reversal and recurrence both seem to violate the Second Law.

By looking at small systems [such as a harmonic oscillator exposed to a temperature gradient, $T(q)) = 1 + \epsilon \tanh(q)$], we found that the phase-space description of such systems is invariably a unidirectional flow "from" a fractal "repellor" "to" a "strange attractor." The attractor and repellor correspond to *velocity* mirror images. The motion forward in time is attractive while its reverse is exponentially *unstable and unobservable.* Fig. 2.10 shows a strange-attractor cross section for the heat-conducting harmonic oscillator with a temperature profile $T(q) = 1 + 0.50 \tanh(q)$ as a plot of penetrating points in the $\zeta = 0$ plane. Just as in Fig. 2.9, the sign of the Lyapunov exponent neither reflects the time-reversibility of the equations of motion nor their inversion symmetry. Both symmetries are *broken* and *missing*.

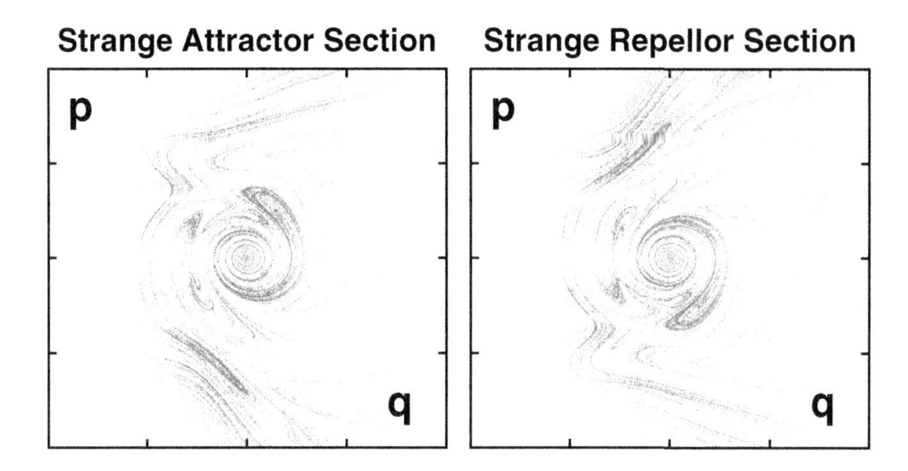

Fig. 2.10 The nonequilibrium harmonic oscillator, with $T(q) = 1 + 0.50\tanh(q)$, shows no symmetry whatsoever in its local largest Lyapunov exponent. Not only does the dynamics depend only on the past (so that there is no symmetry on changing the sign of p). The dynamics must also obey the Second Law of Thermodynamics, so that there is no inversion symmetry through the center of the ($qp0$) cross section. Heat necessarily flows counter to the temperature gradient in accord with the Law.

The Galton Board problem (a particle falling at constant kinetic energy through a periodic array of scatterers) and the conducting oscillator problem (an oscillator conducting heat in the presence of a temperature gradient) both furnish three-dimensional models of nonequilibrium steady states. Both provide ergodic fractal geometry, Lyapunov instability, and the symmetry breaking associated with time reversal and with the Second Law of Thermodynamics. This explanation of the Second Law is identical for many body problems too. My interest in few-body systems was inherited from my earliest work with Berni, dating back to our exploration of one-body cell models for melting, as well as our subsequent few-body thermodynamic studies of disks and spheres.

The Lyapunov *spectrum* (with all n exponents describing the comoving expansion and contraction in an n-dimensional space) characterizes the spatial dependence of instabilities in *all* n of the phase-space directions. It shows that the predominantly negative Lyapunov spectrum describing the condensation of attractive sets onto fractal objects mirrors the predominantly positive spectrum describing the exponentially fast departure of trajectories in the vicinity of the repellor. It is interesting and significant that, despite the time-reversibility of the equations of motion, the stability of the motions forward and backward is qualitatively different. Just as on a roller coaster or a curvy road the passengers' motions are sensitive to both the direction of travel and to the strange-attractor direction imposed by the Second Law.

Caricatures of these problems, carried out here at Livermore, led to this new understanding of the Second Law of Thermodynamics, first described in a 1987 paper (submitted twice, with two different titles!) with Brad Holian and Harald Posch [Holian *et al.* (1987)]. In nonequilibrium steady states the motion forward in time has an attractive fractal distribution in phase space with a phase volume of zero indicating the extreme rarity of the states participating in a steady flow. The time-reversed states, equally rare, make up an inaccessible repellor with a summed-up Lyapunov spectrum which is positive rather than negative. In this way, we found that a time-reversible mechanics, based on Nosé's ideas, provides a clear foundation for an *irreversible* thermodynamics based on *time-reversible* motion equations.

2.9 Conclusion

Berni's research style, beginning with a simple model confirming a "horseback guess," followed by painstaking analyses leading to a clear intuitive description of the work and its significance, has led to rapid progress in understanding phase transformations, nonlinear transport, and aspects of dynamical-systems theory. The 2013 Nobel Prizes in Chemistry rewarded Martin Karplus, Michael Levitt, and Arieh Warshel for applications of *classical* and *thermostated* molecular dynamics to biomedical problems. Berni has always emphasized the need for simplicity and clarity in his work, with an emphasis on words rather than equations and intuitive arguments rather than formal proofs. After more than a half century of research, his way of working seems natural to me and I recognize Berni as a good part of its source. I am looking forward to seeing more of his inspirational work in the years ahead.

2.10 Acknowledgments

My Wife and research colleague, Carol, has contributed so much to this work. She, Berni, and a hundred or so collaborators have my Love and Gratitude. I very much appreciate the efforts of those at the Livermore Laboratory and Brenda Rubenstein and Eric Schwegler in particular, for organizing the celebration of Berni's work and for helping Carol and me to attend it.

Bibliography

Alder, B. J., Hoover, W. G., and Wainwright, T. E. (1963). Cooperative motion of hard disks leading to melting, *Physical Review Letters* **11**, pp. 241–243.

Alder, B. J., Hoover, W. G., and Young, D. A. (1968). Studies in molecular dynamics. v. high-density equation of state and entropy for hard disks and spheres, *The Journal of Chemical Physics* **49**, pp. 3688–3696.

Alder, B. J. and Wainwright, T. E. (1959). Molecular motions, *Scientific American* **201**, pp. 113–126.

Duff, R. E., Gust, W. H., Royce, E. B., Ross, M., Mitchell, A. C., Keeler, R. N., and Hoover, W. G. (1968). *Shockwave Studies in Condensed Media* (Gordon and Breach, New York).

Holian, B. L., Hoover, W. G., Moran, B., and Straub, G. K. (1980). Shockwave structure *via* nonequilibrium molecular dynamics and Navier-Stokes continuum mechanics, *Physical Review A* **22**, pp. 2798–2808.

Holian, B. L., Hoover, W. G., and Posch, H. A. (1987). Second-law irreversibility of reversible mechanical systems: Resolution of Loschmidt's paradox: The origin of irreversible behavior in reversible atomistic dynamics, *Physical Review Letters* **59**, pp. 10–13.

Hoover, W. G. (1985). Canonical dynamics: Equilibrium phase-space distributions, *Physical Review A* **31**, pp. 1695–1697.

Hoover, W. G. and Ashurst, W. T. (1975). *Nonequilibrium Molecular Dynamics* (Academic Press, New York), pp. 1–51.

Hoover, W. G., De Groot, A. J., Hoover, C. G., Stowers, I. F., Kawai, T., Holian, B. L., Boku, T., Ihara, S., and Belak, J. (1990). Large-scale elastic-plastic indentation simulations *via* molecular dynamics, *Physical Review A* **42**, pp. 5844–5853.

Hoover, W. G. and Holian, B. L. (1996). Kinetic moments method for the canonical ensemble distribution, *Physics Letters A* **211**, pp. 253–257.

Hoover, W. G., Hoover, C. G., and Sprott, J. C. (2015a). Nonequilibrium systems: Hard disks and harmonic oscillators near and far from equilibrium, *Molecular Simulation*, but see instead `arXiv:1507.08302`.

Hoover, W. G., Hoover, C. G., and Uribe, F. J. (2010). Flexible macroscopic models for dense-fluid shockwaves: Partitioning heat and work; delaying stress and heat flux; two-temperature thermal relaxation, `arViv:1005.1525`.

Hoover, W. G., Hoover, N. E., and Hanson, K. (1979). Exact hard-disk free volumes, *The Journal of Chemical Physics* **70**, pp. 1837–1844.

Hoover, W. G. and Ree, F. H. (1968). Melting and communal entropy for hard spheres, *The Journal of Chemical Physics* **49**, pp. 3609–3617.

Hoover, W. G., Sprott, J. C., and Hoover, C. G. (2016). Ergodicity of a singly-thermostated harmonic oscillator, *Communications in Nonlinear Science and Numerical Simulation* , pp. 234–240, `arXiv:1504.07654`.

Hoover, W. G., Sprott, J. C., and Patra, P. K. (2015b). Deterministic time-reversible thermostats: Chaos, ergodicity, and the zeroth law of thermodynamics, *Molecular Physics* **113**, pp. 2863–2872, `arXiv:1501.03875`.

Hoover, W. G., Sprott, J. C., and Patra, P. K. (2015c). Ergodic time-reversible chaos for Gibbs' canonical oscillator, *Physics Letters A*, `arXiv:1503.06749`.

Levesque, D., Verlet, L., and Kürkijarvi, J. (1973). Computer 'experiments' on classical fluids. iv. transport properties and time-correlation functions of the Lennard-Jones liquid near its triple point, *Physical Review A* **7**, pp. 1690–1700.

Nosé, S. (1984a). A molecular dynamics method for simulations in the canonical ensemble, *Molecular Physics* **52**, pp. 255–268.

Nosé, S. (1984b). A unified formulation of the constant temperature molecular dynamics methods, *The Journal of Chemical Physics* **81**, pp. 511–519.

Rahman, A. (1964). Correlations in the motion of atoms in liquid argon, *Physical Review A* **136**, pp. 405–411.

Chapter 3

Why Non-Equilibrium is Different

J. Robert Dorfman, Theodore R. Kirkpatrick, and Jan V. Sengers

Institute for Physical Science and Technology University of Maryland, College Park, MD 20742, USA

Abstract

The 1970 paper, "Decay of the Velocity Correlation Function" [Phys. Rev. **A1**, 18 (1970), see also Phys. Rev. Lett. **18**, 988, (1967)] by Berni Alder and Tom Wainwright, demonstrated, by means of computer simulations, that the velocity autocorrelation function for a particle moving diffusively in a gas of hard disks decays algebraically in time as t^{-1}, and as $t^{-3/2}$ for a gas of hard spheres. These decays appear in non-equilibrium fluids and have no counterpart in fluids in thermodynamic equilibrium. The work of Alder and Wainwright stimulated theorists to find explanations for these "long time tails" using kinetic theory or a mesoscopic mode-coupling theory. This paper has had a profound influence on our understanding of the non-equilibrium properties of fluid systems. Here we discuss the kinetic origins of the long time tails, the microscopic foundations of mode-coupling theory, and the implications of these results for the physics of fluids. We also mention applications of the long time tails and mode-coupling theory to other, seemingly unrelated, fields of physics. We are honored to dedicate this short review to Berni Alder on the occasion of his 90th birthday!

3.1 Divergences in Non-Equilibrium Virial Expansions

N. N. Bogoliubov [Bogoliubov (1949)], by means of functional assumption methods, and later M. S. Green [Green (1956)] and E. G. D. Cohen [Cohen (1962)], using cluster expansion methods, independently solved the outstanding problem in the non-equilibrium statistical mechanics of gases at the time, namely, to extend the Boltzmann transport equation to dense gases as a power series expansion in the density of the gas. These authors were able to formulate a generalized Boltzmann equation for monatomic gases with short ranged central potentials, in the form of a virial expansion of the collision operator whose successive terms involved the dynamics of isolated groups of two, three, four, etc., particles interacting amongst themselves. The generalized Boltzmann equation was written by these authors as [Cohen (1963); Green and Piccirelli (1963)]

$$\frac{\partial f(\mathbf{r},\mathbf{v},\mathbf{t})}{\partial t} + \mathbf{v}\cdot\nabla_{\mathbf{r}} f((\mathbf{r},\mathbf{v},\mathbf{t}) = J_2(f,f) + J_3(f,f,f) + \cdots \quad (3.1)$$

Here, $f((\mathbf{r},\mathbf{v},\mathbf{t})$ is the single particle distribution function for finding particles at position $\mathbf{r}$ with velocity $\mathbf{v}$ at time t. The collision operator J_2 is the usual Boltzmann, binary collision operator, while the J_j are collision operators determined by the dynamical events taking place among an isolated group of j particles.

At roughly the same time as the problem of generalizing the Boltzmann equation to higher densities was being addressed, M. S. Green [Green (1952, 1954)] and R. Kubo [Kubo (1957)] independently developed a general method for expressing the transport coefficients appearing in the linearized equations of fluid dynamics in terms of time integrals of equilibrium time correlation functions of microscopic currents. These expressions have the general form

$$\xi(n,T) = \int_0^\infty dt\, \langle j_\xi(0) j_\xi(t)\rangle_{eq}\,. \quad (3.2)$$

Here, $\xi(n,T)$ is a transport coefficient such as the coefficient of shear viscosity, thermal conductivity, *etc.*, at fluid density n and temperature, T, the brackets denote an equilibrium ensemble average, and $j_\xi(t)$ is the value of an associated microscopic current at some time t. An example that will be important for our discussion is the case of tagged-particle diffusion whereby one particle in a gas of mechanically identical particles has some non-mechanical tag that enables one to follow its diffusion in the gas. For this case, the diffusion coefficient D is given by the Green-Kubo formula:

$$D(n,T) = \int_0^\infty dt \, \langle v_x(0) v_x(t) \rangle_{eq} \,, \tag{3.3}$$

where $v_x(t)$ is the $x-$component of the velocity of the tagged particle.[1]

From the density expansion of the collision operator or by an equivalent cluster expansion of the time correlation function expressions, one can, at least in principle, obtain expressions for transport coefficients of the gas as power series in the density, similar to the virial expansions for the equilibrium properties of the same gas. This parallel development indicated the existence of a "super statistical mechanics," whereby both equilibrium and non-equilibrium properties of a gas can be expressed in the form of virial, or power series, expansions in the density of the gas, obtained by means of almost identical cluster expansion methods. However, it quickly became clear that this parallelism was purely illusory: the non-equilibrium properties of a gas have almost nothing in common with its equilibrium

[1]We mention that transport coefficients, characterizing non-equilibrium flows are, in the Green-Kubo formalism, expressed in terms of time correlation functions measured in an equilibrium ensemble. This is consistent with Onsager's assumption that the final stages of the relaxation of microscopic fluctuations about an equilibrium state can be described by macroscopic hydrodynamic equations. Another example occurs in the treatment of dynamic light scattering by fluids in equilibrium.

properties. The first indication of this situation appeared in 1965 when Dorfman and Cohen [Dorfman and Cohen (1965)], among others [Brush (1972)], discovered that almost every term in the non-equilibrium virial expansions diverges!

The differences between the equilibrium and non-equilibrium virial expansions have their origins in the type of correlations upon which the virial coefficients depend. The equilibrium virial coefficients depend only upon static correlations between a fixed number of particles in contrast to the non-equilibrium virial coefficients which depend mainly, if not exclusively, upon dynamical correlations produced by sequences of collisions taking place between a fixed number of particles. To be explicit, the equilibrium virial expansion for the pressure, $p(n, T)$, of a gas at number density n and at temperature T, and the non-equilibrium virial expansion for the transport coefficient, ξ, of a gas at local number density n and at local temperature T are given by

$$\frac{p(n,T)}{nk_BT} = 1 + nb_1(T) + n^2b_2(T) + \cdots, \tag{3.4}$$

$$\frac{\xi(n,T)}{\xi_0(T)} = 1 + n\sigma^d a_1^{(\xi)}(T) + (n\sigma^d)^2 a_2^{(\xi)}(T) + \cdots. \tag{3.5}$$

with k_B Boltzmann's constant. Here, $\xi_0(T)$ is the low density value of the transport coefficient as determined by the Boltzmann equation for the gas and σ is the range of the intermolecular force. The coefficients $b_{j-1}(T)$ are determined by static correlations among j interacting particles. The range of these static correlations is at most $j\sigma$. In contrast, the non-equilibrium virial coefficient $a_{j-2}^{(\xi)}$ depends upon correlated sequences of collisions taking place among a group of j particles in infinite space and over an arbitrarily long time interval between the first and final collision of the sequence. For the systems under discussion here, all the equilibrium virial coefficients, b_j, are finite and of order $(\sigma^d)^j$, where d is the spatial dimension of the system. However, all but the first few non-equilibrium virial coefficients diverge! For two-dimensional systems, the coefficients

$a_1^{(\xi)}, a_2^{(\xi)}, \ldots$, all diverge[Dorfman and Cohen (1967)]. For three-dimensional systems, the coefficients a_2 and higher all diverge.

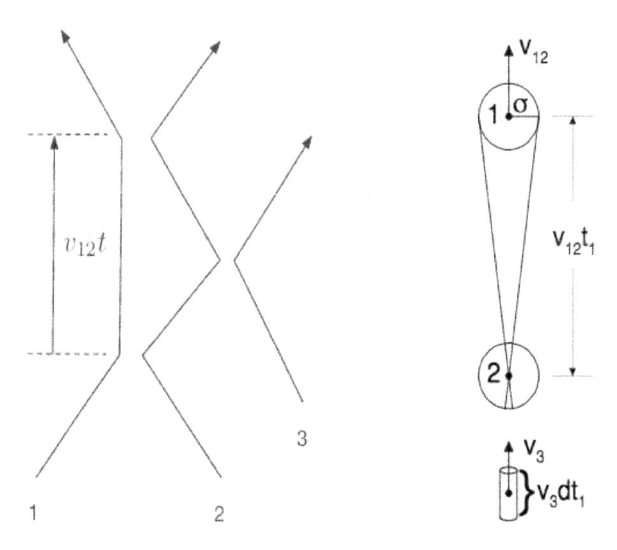

Fig. 3.1 Three particle recollision event: Particles 1 and 2 collide at some initial time, and due to an intermediate $(2, 3)$ collision, particles 1 and 2 collide at a time t later. The second figure is set in the rest frame of particle 2, and shows the solid angle into which particle 2 must be scattered by particle 3 to collide with particle 1 at time t. Here v_{12} is the relative velocity of particles 1 and 2 after the first collision.

The origin of these divergent coefficients can be easily understood by considering the coefficient $a_1^{(\xi)}$, for example. In Figure 3.1, we illustrate one of the three-particle, correlated collision sequences that contribute to this coefficient [Green and Piccirelli (1963); Dorfman and Cohen (1967)]. In this recollision event, particles 1 and 2 collide at some initial instant, then later, particle 3 collides with particle 2, in such a way that particles 1 and 2 collide again after a time interval t between the first and last collisions between these two particles. The sequences take place in infinite space and over arbitrarily large times, t. As illustrated in the Figure, the dynamics is controlled by the solid angle into which particle 2 must be scattered when particle 3 hits it. The phase space region available for particle 3 to cause the $(1, 2)$ re-collision between time t and $t+dt$ is proportional to the solid angle and is of the order $(\sigma/vt)^{d-1}dt$. The coefficient

a_1 is determined by the integration of this region over all possible time intervals t, and is clearly logarithmically divergent for $d = 2$. The coefficient $a_1^{(\xi)}$ is finite for $d = 3$, but the next coefficient, $a_2^{(\xi)}$, is logarithmically divergent for three-dimensional systems for similar reasons, and all higher coefficients diverge also, as powers of the upper limit on the time integral which can be arbitrarily large. Thus we can identify the essential difference between equilibrium and non-equilibrium properties of gases: non-equilibrium processes are due, among other things, to dynamical processes that can take place over large spatial distances and over large times. These processes cause long range and long time correlations among the particles in the gas that are absent in equilibrium, except perhaps at critical points, and even then, are of a qualitatively different origin. We are now faced with another problem. The results of Bogoliubov, Green, and Cohen are incomplete - their virial series are useless for descriptions of processes that take place over times long compared to some microscopic time due to the long time divergences in the terms in the virial series.

3.2 The Ring Resummation

It is clear what is causing the divergences in the non-equilibrium expansions. A collective effect, mean free path damping of trajectories, has been ignored when deriving the virial expansions. We argued above that the virial coefficients depend on the dynamics of isolated groups of a fixed number of particles, and the time between any two collisions in the troublesome correlated collision sequences can be arbitrarily large. This is clearly unphysical. In a real gas, particles cannot travel arbitrarily long distances between collisions without another particle interrupting the motion of the particles by colliding with one of them. That is to say, the typical distance between collisions is a mean free path which in turn depends upon the gas density and temperature. The probability of a particle moving a certain distance is exponentially damped as the distance of travel becomes larger than a few mean free path lengths. In essence, by insisting that the collision operator or that the time correlation expressions be expanded in a power series in the gas density, one has taken what should be an exponential damping and expressed the exponential as a power series. Thus non-equilibrium virial expansions are very misleading since they are the equivalent of writing

$$e^{-nt} = 1 - nt + \frac{1}{2}(nt)^2 + \cdots , \tag{3.6}$$

and trying to determine the behavior of the exponential by examining individual terms on the right-hand side of its power series expansion. It is clear that a more physical representation of the generalized collision operator or of the time correlation function expressions should be obtained by summing the most divergent terms in the virial expansions and using the resummed expression, not the virial expansions. This resummation was first carried out by K. Kawasaki and I. Oppenheim in 1965 [Kawasaki and Oppenheim (1965)]. They expressed the most divergent terms in the virial expansions as ring events and were able to resum these most divergent terms and to obtain expressions for transport coefficients that should be well behaved, in contrast to the virial expansion representations.[2] For three-dimensional systems, one can use the resummed expressions for the transport coefficients to show that the logarithmic divergence in the virial expansion is replaced by a logarithmic term in the density that results from including the mean free path damping in the relevant collision integrals. Thus for gases in three dimensions the first few terms in the density expansion of the transport coefficients are

$$\frac{\xi(n,T)}{\xi_0(T)} = 1 + a_1^{(\xi)} n\sigma^3 + a_{2,ln}^{(\xi)} (n\sigma^3)^2 \ln n\sigma^3 + a_{2,n}^{(\xi)} (n\sigma^3)^2 + \cdots . \tag{3.7}$$

[2]It is worth pointing out that nothing like this has to be done for equilibrium virial expansion if the gas is composed of particles interacting with short range forces. However, if the gas is composed of particles interacting with infinite range Coulomb potentials, a similar ring summation is necessary even in equilibrium.

The coefficient of the linear term, $a_1^{(\xi)}$, had already been calculated by Sengers [Sengers (1967)] for hard spheres, based on the analysis of this three-body collision integral by S. T. Choh and G. E. Uhlenbeck [Choh and Uhlenbeck (1958)] and by Green [Green (1964)]. Also, for hard spheres the coefficient, $a_{2,ln}^{(\xi)}$, of the logarithmic term has been calculated [Kamgar-Parsi and Sengers (1983)], and estimates have been made of the coefficient $a_{2,n}^{(\xi)}$ using the Enskog theory. In Figure 3.2, we show the comparison of the theoretical and computer results for the coefficients of self-diffusion, shear viscosity, and thermal conductivity for a moderately dense gas of hard spheres [Dorfman *et al.* (1994)]. The agreement is quite good despite the fact that the coefficient $a_{2,n}^{(\xi)}$ can only be estimated for reasons that will become clear below.[3]

The non-analytic terms in a density expansion of a transport coefficient have also been considered in quantum systems. Indeed, very early on J. S. Langer and T. Neal [Langer and Neal (1966)], motivated by the above classical work, pointed out that logarithmic terms appear in the electrical conductivity in disordered electronic systems. It can be argued that this sort of calculation, basically a quantum Lorentz gas, is also relevant for the electron mobility, μ, of excess electrons in liquid helium. In this case the dimensionless density expansion parameter, $\chi = 4na_s^2\lambda$, involves the thermal de Broglie wavelength, $\lambda = (2\pi^2\hbar^2/mk_BT)^{1/2}$, the density of helium atoms, n, and the $s-$wave scattering length, a_s. Here, $\hbar$ is Planck's constant and m is the electron mass. Wysokinski, Park, Belitz and Kirkpatrick [Wysokinski *et al.* (1994, 1995)] have computed exactly μ up to and including terms of $O(\chi^2)$ and obtained,

$$\mu/\mu_B = 1 + \mu_1\chi + \mu_{2ln}\chi^2 ln\chi + \mu_2\chi^2 + o(\chi^2). \tag{3.8}$$

[3]To anticipate this discussion, we mention that the value of this coefficient depends upon the long time behavior of the relevant time correlation functions or upon a good guess at a lower cut-off of the time integrations.

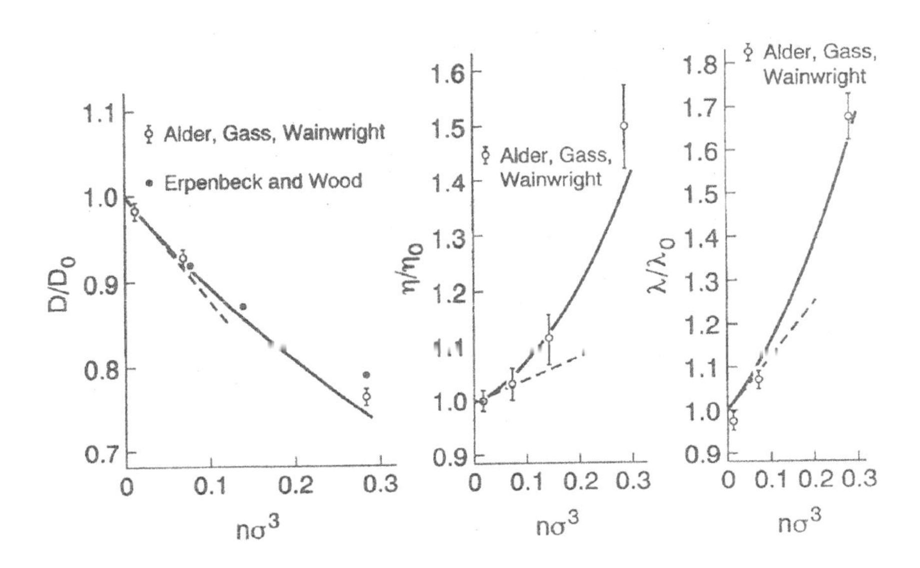

Fig. 3.2 Comparison of the theoretical expressions, Eq.(3.7) for the transport coefficient of self-diffusion, D, the coefficient of shear viscosity, η, and the thermal conductivity, λ, for a gas of hard spheres with the results of molecular dynamics [Dorfman *et al.* (1994); Alder *et al.* (1970); Erpenbeck and Wood (1991)]. The dashed lines correspond to keeping only the first two terms in this expansion. The coefficient $a_{2,n}^{(\xi)}$ is estimated using the Enskog theory.

Here, μ_B is the Boltzmann equation value for μ, and $\mu_1 = -\pi^{3/2}/6$, $\mu_{2ln} = (\pi^2 - 4)/32$, and $\mu_2 = 0.236\ldots$. Adams *et al.* [Adams *et al.* (1992)] have concluded that existing experiments give very good agreement with the value of the conductivity given by Eq. (3.8).

To test experimentally the presence or absence of the logarithmic term, Wysokinski *et al.* defined the function,

$$f(\chi) = [\mu/\mu_B - 1 - \mu_1\chi]/\chi^2. \tag{3.9}$$

Theoretically,

$$f(\chi) = \mu_{2ln} ln\chi + \mu_2 \pm 2\pi^{1/2}\chi, \tag{3.10}$$

where the last term is an estimate of the $O(\chi^3)$ contribution to μ. In Figure 3.3 [Wysokinski *et al.* (1994, 1995)], the theoretical prediction is shown for $0 < \chi < 0.7$ together with experimental data [Schwartz (1980)]. The error bars shown assume a total error of 3% in μ/μ_B and 4% in χ. To illustrate the effect of the logarithmic term, the figure also shows what the theoretical prediction would be if μ_{2ln} in Eq.(3.8) were zero.

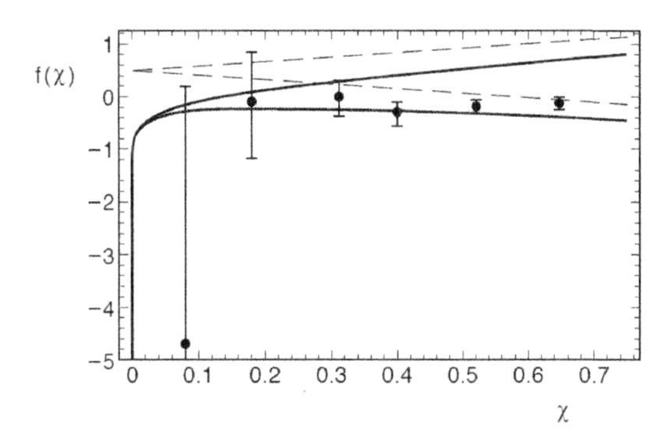

Fig. 3.3 The reduced mobility f, as defined in Eq.(3.9), *vs.* the density parameter χ. The theoretical prediction of Wysokinski, Park, Belitz, and Kirkpatrick is for f to lie between the two solid curves [Wysokinski *et al.* (1994)]. The experimental data are from Fig. 9 of Ref. [Schwartz (1980)] with error bars estimated as in the two possible forms of Eq.(10). The broken lines show what the theoretical prediction would be in the absence of the logarithmic term in the density expansion.

Following the discovery that logarithmic terms must appear in non-equilibrium density expansions, there were strong indications that something was still amiss in the kinetic theory for transport coefficients. In 1966, R. Goldman [Goldman (1966)] argued that the resummed expressions for transport coefficients contain time integrals of functions with power-law decays. He identified the leading power as $t^{-3/2}$ for long times, for three-dimensional systems. In 1968, Y. Pomeau [Pomeau (1971)] argued that, for two-dimensional systems, the Kawasaki-Oppenheim expressions still diverge as time integrals of functions that decay as t^{-1} for large times.

This was the situation just before the work of Alder and Wainwright on the velocity auto-correlation function became known, and before the appearance in 1970 of their paper in Physical Review which stimulated so much work in non-equilibrium statistical mechanics, and continues to reverberate even now with new and unexpected applications.

3.3 The Alder-Wainwright Paper of 1970: Long Time Tails

The papers by Alder and Wainwright in Physical Review Letters in 1967 [Alder and Wainwright (1967)], and most especially, that in Physical Review in 1970 [Alder and Wainwright (1970)], provided the spark that ignited the imaginations of those of us working in kinetic theory. They considered gases of hard spheres or of hard disks at moderate densities and by means of computer simulated molecular dynamics, obtained the velocity correlation function $\langle v_x(0)v_x(t)\rangle \,/\, \langle v_x^2\rangle$ for a range of times, scaled with the appropriate mean free time, t_m, between collisions. Their results provided convincing evidence that over a range of times, roughly $10 \leq s = t/t_m \leq 30$, the velocity autocorrelation functions decay as

$$\frac{\langle v_x(0)v_x(s)\rangle_{eq}}{\langle v_x^2\rangle_{eq}} \simeq \alpha_D^{(d)}(n)s^{-d/2}. \tag{3.11}$$

Here $\alpha_D^{(d)}(n)$ is a numerical coefficient that depends on the density and the spatial dimension of the gas. The subscript D indicates that the time correlation function is the one needed for the coefficient of self- or tagged particle diffusion through Eq. (3.3). Figure 3.4 shows their results for the three-dimensional case.

Stimulated by these computer results and following theoretical arguments of Goldman and Pomeau, Dorfman and Cohen [Dorfman and Cohen (1970, 1972)] were able to show that these algebraic decays can be explained both qualitatively and quantitatively by kinetic theory. They evaluated the Kawasaki-Oppenheim ring summation, but in order to obtain results appropriate for the densities studied by Alder and Wainwright, they extended the summation result to higher densities by means of the Enskog theory for dense hard ball gases [Dorfman and Cohen (1975)]. At the same time, Ernst, Hauge, and van Leeuwen [Ernst *et al.* (1970)] provided a mesoscopic argument for these

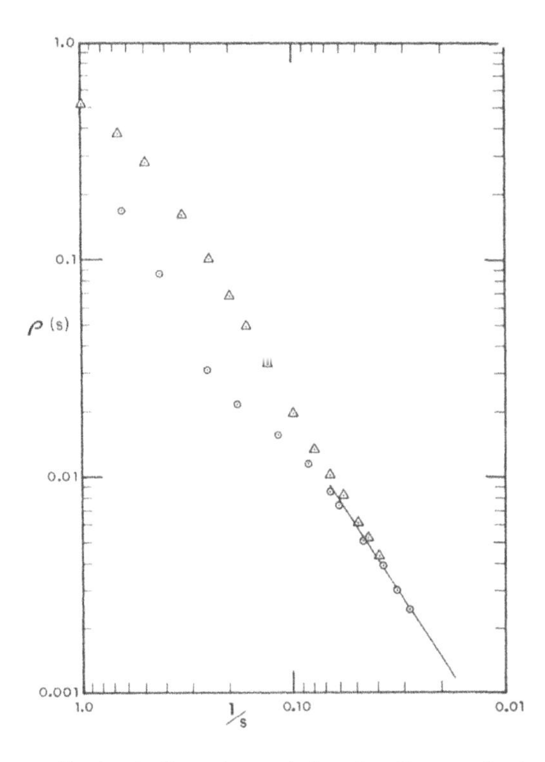

Fig. 3.4 The normalized velocity autocorrelation function as obtained by molecular dynamics (triangles) for a gas of 500 hard spheres [Alder and Wainwright (1970)]. The circles are results obtained using a hydrodynamical model developed by Alder and Wainwright to explain their results.

algebraic decays, or as they are called now, *long time tails.* The expression for the coefficient $\alpha_D^{(d)}(n)$ will serve to illustrate a general feature of the theoretical explanation of the long time tails,

$$\alpha_D^{(d)}(n) = c_d\left[(D+\nu)t_m\right]^{-d/2}. \tag{3.12}$$

Here, c_d is a numerical coefficient and proportional to n^{2-d}, D is the coefficient of self-diffusion, $\nu = \eta/\rho$ is the kinematic viscosity, η is the coefficient of shear viscosity and ρ is the mass density of the fluid. The comparison of the kinetic theory results, using the Enskog theory for the transport coefficients with the results of Alder and Wainwright is illustrated in Figure 3.5 [Dorfman and Cohen (1970)]. This provides conclusive proof that the Alder-Wainwright results can be explained by kinetic theory when the contributions of the most divergent terms are taken into account.

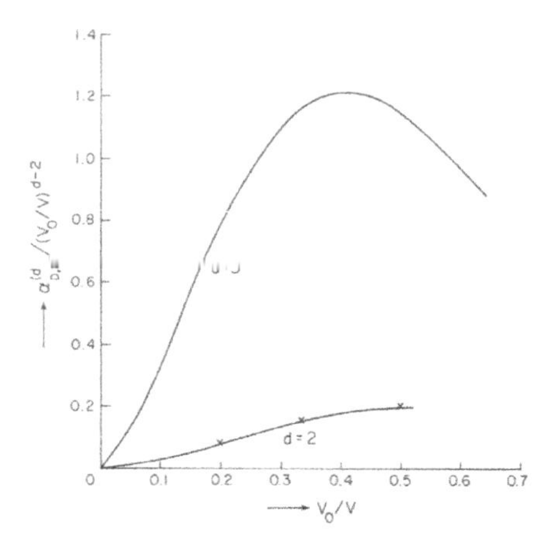

Fig. 3.5 The solid curves are the theoretical results obtained by Dorfman and Cohen for the coefficients, $\alpha_{D,E}^{(d)}$ appearing in Eq. (3.12) for the velocity autocorrelation function, using Enskog theory values for the transport coefficients [Dorfman and Cohen (1970)]. The crosses represent values obtained from molecular dynamics by Alder and Wainwright, for two-dimensional systems [Alder and Wainwright (1970)]. Here V/V_0 is the ratio of the volume of the system to that at close packing of the disks or spheres.

In Figure 3.6 we show later results of Wood and Erpenbeck [Wood and Erpenbeck (1976)] confirming those of Alder and Wainwright and they compared their results with theoretical results including finite size effects.

It is important to note that $\alpha_D^{(d)}$ depends upon the sum of two transport coefficients, in this case the coefficient of self-diffusion and the kinematic viscosity. This is an indication of the fact that the underlying microscopic processes generating the tails are the coupling of microscopic hydrodynamic modes that exist as fluctuations in fluids and are detected in dynamic light scattering experiments.[4] The dynamical events taking place in

[4]The Rayleigh and Brillouin peaks seen in dynamic light scattering by an equilibrium fluid are due to microscopic heat and sound modes appearing as fluctuations about equilibrium in the fluid.

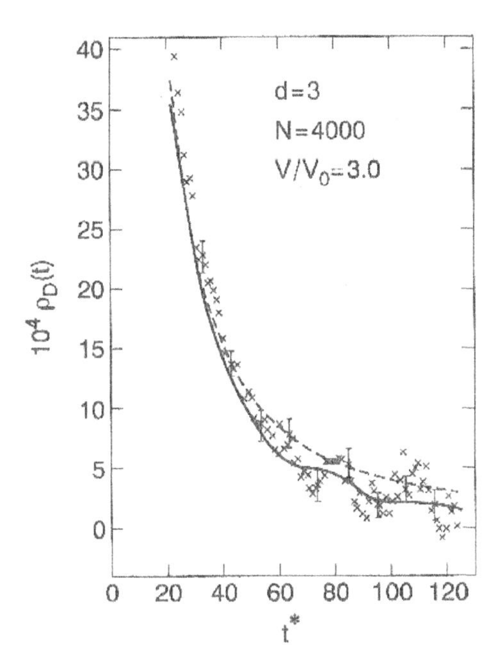

Fig. 3.6 Results of Wood and Erpenbeck for the velocity autocorrelation function for a gas of 4000 hard spheres at a volume of three times the close packing volume [Wood and Erpenbeck (1976)]. Here, t^* is the time, measured in mean free times. The crosses are computer results, the dashed line is that given by Eq.(3.11), and the solid line represents a complete evaluation of the mode coupling formula with all modes taken into account and finite size corrections included.

the gas generate both the modes and their coupling. A simple example will illustrate the point. A somewhat oversimplified picture of a renormalized recollision illustrated in Figure 3.1 is shown in Figure 3.7 [Dorfman (1981)]. Two particles collide at some instant of time, then undergo an arbitrary number of intermediate collisions before recolliding at time t. One can think of the motions of the two particles after their first collision as random walks that cross at time t. If we sit on one of the particles, we can imagine that the recollision is a random walk that returns to the origin. A standard calculation in random walk theory shows that the probability of a return to the origin after a time interval t is proportional to $(1/t)^{d/2}$. This time dependence is exactly that of the long time tails, and the random walks represent hydrodynamic processes such as diffusion that are coupled by the initial and final collisions.[5]

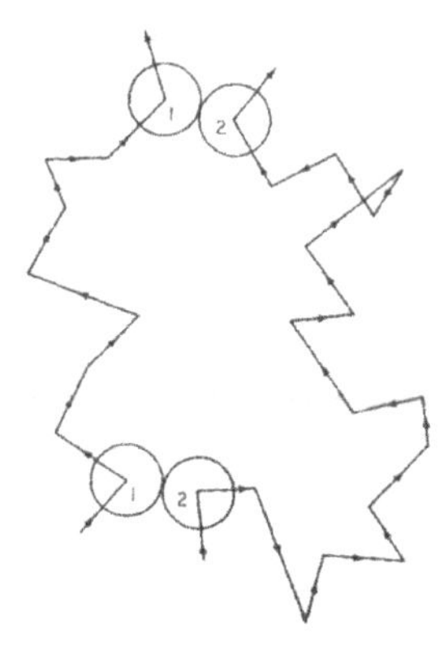

Fig. 3.7 A schematic version of a renormalized recollision sequence. Two particles collide and then each of them undergoes a random walk produced by collisions with other particles before they collide again after a time t.

We thus have, when this is all worked out properly, a microscopic derivation of *mode-coupling theory*, already known from the work of L. P. Kadanoff and J. Swift [Kadanoff and Swift (1966)] and of Kawasaki [Kawasaki (1970)] on the behavior of transport coefficients near the critical point of a phase transition. In fact, the Kadanoff-Swift results are exactly the combined result of the long time tail processes with the behavior of thermodynamic properties near a critical point. We also mention that the transport coefficients appearing in the expression for $\alpha_D^{(d)}$ have to be treated with some care. They cannot be the full transport coefficients since those are determined by the long time behavior of correlation functions. Instead, over the time of

[5]For tagged particle diffusion, only one of the initial colliding pair is followed, while the other particles in the collision sequences can be any other particles in the fluid. The tagged particle motion is represented by the appearance of the diffusion coefficient in the long time tail result, Eq. (3.12), while the motions of the other particles in the sequence are represented by the viscous mode contribution to this formula.

the Alder-Wainwright studies, these are to be seen as short time contributions, thus accounting for the success of using the Enskog expressions for the transport coefficients when comparing the computer results with those from kinetic theory.

3.4 Consequences of the Long Time Tails for Hydrodynamics

The algebraic time decays of the time correlations and the existence of generic long range correlations in non-equilibrium systems have immediate consequences for microscopic derivations of the Navier Stokes and higher-order hydrodynamic equations. The most immediate of these is that for gases in two dimensions the time correlation function expressions for transport coefficients diverge logarithmically with the upper limit of the time integrals in the Green-Kubo formulas. For three-dimensional systems, the Navier Stokes transport coefficients are finite, but transport coefficients in higher-order equations, such as the Burnett equations diverge [Dorfman and Cohen (1970); Ernst *et al.* (1970); Ernst and Dorfman (1975); Pomeau and Resibois (1975)]. We are therefore faced with the fact that our microscopic derivations of the fluid dynamics equations have divergence problems. A number of studies have been carried out in order to determine a more correct form of these equations, free of divergence problems. The results are complicated and depend to a certain extent on the transport process. For example, for two-dimensional viscous flows, one finds that Newton's law of viscous friction must be modified by the addition of non-linear logarithmic terms in the velocity gradients. For three-dimensional systems, there is a non-analytic correction to Newton's law. That is, the off-diagonal terms of the pressure tensor, P_{xy}, for example have the form [Ernst *et al.* (1978); Onuki (1979)]

$$\begin{aligned} P_{xy}^{(2)} &\simeq -\tilde{\eta} X + aX \ln X + \cdots , \\ P_{xy}^{(3)} &\simeq -\eta X + bX|X|^{1/2} + \cdots , \\ X &= \frac{\partial u_x(y)}{\partial y}. \end{aligned} \tag{3.13}$$

Here, $u_x(y)$ is the component of the fluid velocity, $\mathbf{u}$, that is a function of the coordinate in a perpendicular direction, as is appropriate for shear flow. We see that for two-dimensional systems, viscous flow is inherently nonlinear, since a coefficient of shear viscosity defined by the limit $\lim_{X\to 0} P_{xy}/X$ does not exist. For three-dimensional systems, the corrections to Newton's law are non-analytic; in this case, a fractional power of the velocity gradient appears. Physically, a finite shear rate weakens or makes shorter range the correlations that cause the divergence problems.

The same considerations have also been applied to the case of a stationary temperature gradient. Surprisingly, this case is very different. A finite ∇T does not fix the divergence problem in the two-dimensional heat conductivity, nor does it lead to non-analytic terms in the three-dimensional heat flux [Kirkpatrick (1981); Kirkpatrick and Dorfman (2015)]. This in turn implies that correlations in a non-equilibrium system with a temperature gradient are of longer range, and more robust, than a fluid with a velocity gradient. This observation is intimately tied to the striking results discussed in the next section.

For three-dimensional systems the dispersion relation for sound propagation in a gas also has a non-analytic form [Pomeau (1973); Ernst and Dorfman (1975)],

$$\omega(k) = \pm ick + \frac{1}{2}\Gamma k^2 + Ak^{5/2} + \cdots , \tag{3.14}$$

where $\omega(k)$ is the frequency of sound as a function of the wave number, k, c is the velocity of sound and Γ is the sound damping constant. There are also an infinite number of terms between k^2 and k^3, only the first of which is given here. There is, indeed, some experimental evidence for the appearance of the $k^{5/2}$ term in this dispersion relation as seen from neutron scattering studies on liquid sodium [Morkel and Gronemeyer (1988)]. It is possible to analyze the neutron scattering data in order to obtain values of the frequency dependence of the Fourier transform, $Z(\omega)$, of the velocity correlation function as a function of the frequency, ω. The long time tail in this function would then be seen as a dependence of the Fourier transform on $\omega^{1/2}$. The results of Morkel *et al.* are illustrated in Figure 3.8. The square root dependence is clear and the data are in good agreement with the theory.

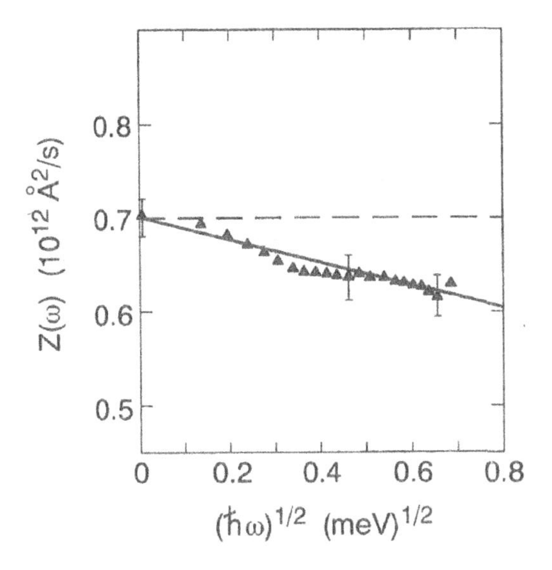

Fig. 3.8 The Fourier transform of the velocity autocorrelation function, $Z(\omega)$, as a function of the square root of the frequency, ω, for atoms in liquid sodium as obtained from neutron scattering experiments (triangles). The solid line is the theoretical result including mode-coupling effects, while the dashed line omits them [Morkel and Gronemeyer (1988)].

In general, very little is known about the complete structure of the hydrodynamic equations, especially for two-dimensional systems. Non-analytic terms, finite size effects, branch point structures, and so on seem to be present. The only redeeming feature of all of this is that these complications do not appreciably distort the results obtained by using ordinary Navier Stokes hydrodynamics, even if, for two dimensions we can only give approximate results for the transport coefficients that appear in them.

3.5 Non-Equilibrium Steady States

Very dramatic deviations from equilibrium behavior due to mode-coupling effects causing long range spatial correlations can be found in the properties of fluids maintained in non-equilibrium stationary states. The first striking example of this difference was discovered by Kirkpatrick [Kirkpatrick (1981)], described in his doctoral dissertation and in a subsequent series of papers by Kirkpatrick, Cohen and Dorfman [Kirkpatrick *et al.* (1982a,b,c)]. Confirmation of this work was obtained by Sengers and co-workers in a series of light scattering experiments on a fluid maintained in a steady state with a fixed temperature gradient. As we noted above the structure factor for an equilibrium fluid is, for small wave numbers, characterized by a central Rayleigh peak and two Brillouin peaks on either side of the central peak. All of this changes when a constant temperature gradient is imposed on the system. Most dramatic of these effects is the enhancement of the central peak by orders of magnitude, an enhancement due to the long range spatial correlations in a non-equilibrium fluid. When a temperature gradient is imposed on a fluid, the central peak of structure factor, $S_{neq}(t, \mathbf{k})$, for a simple fluid, is given for small wave numbers k as a function of time, t, by

$$\begin{aligned} S_{neq}(t, \mathbf{k}) &= S_0 \left[(1 + A_T)\, e^{-D_T k^2 t} - A_\nu e^{-\nu k^2 t} \right], \\ A_T &= \frac{c_P}{T\left(\nu^2 - D_T^2\right)} \left(\frac{\nu}{D_T} \right) \frac{\left(\hat{\mathbf{k}}_\perp \cdot \nabla T \right)^2}{k^4}, \\ A_\nu &= \frac{c_P}{T\left(\nu^2 - D_T^2\right)} \frac{\left(\hat{\mathbf{k}}_\perp \cdot \nabla T \right)^2}{k^4}. \end{aligned} \tag{3.15}$$

Here S_0 measures the intensity of the thermal fluctuations when the fluid is in equilibrium, c_P is the specific heat capacity at constant pressure, ν, D_T are the coefficients of kinematic viscosity and of thermal diffusivity, respectively, and $\hat{\mathbf{k}}_\perp$ is a unit vector in a direction perpendicular to that of the wave vector, $\mathbf{k}$. It is important to note the inverse fourth power of the wave number appearing in the coefficients A_T, A_ν, and the proportionality to the square of the component of the temperature gradient in a direction perpendicular to that of the wave vector. The strong dependence on the wave number indicates quite clearly that these effects are due to the long range nature of the spatial correlations in the fluid, and these terms vanish for zero temperature gradient. All the thermodynamic and transport coefficients are known for toluene, for example, so that a direct comparison of theory and experiment can be carried out, as was done by Sengers and co-workers [Segrè *et al.* (1992)]. The results are given in Figure 3.9. The agreement of theory and experiment is excellent.

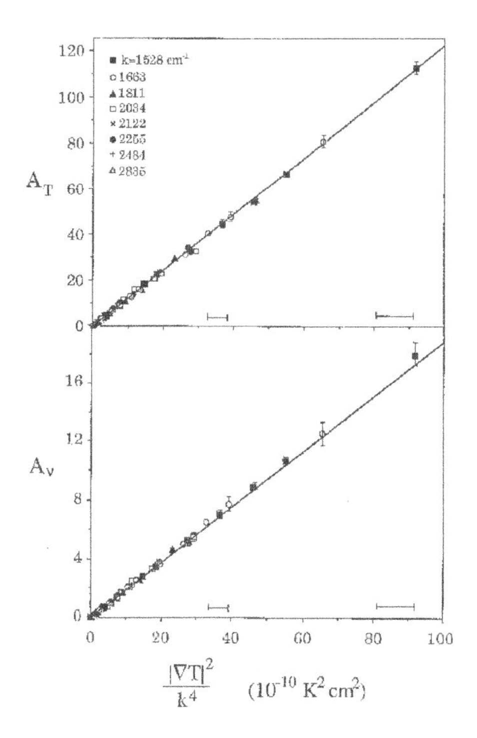

Fig. 3.9 The coefficients A_T and A_ν as a function of wave number as measured in light scattering experiments by Sengers *et al.* [Segrè *et al.* (1992)]. The solid lines are theoretical values with no adjustable parameters.

Generally, long-ranged fluctuations will also induce a so-called Casimir force in a confined fluid [Kardar and Golestanian (1999)]. A well known example is the Casimir effect due to critical fluctuations in equilibrium fluids [Fisher and de Gennes (1978); Krech (1994)]. Critical fluctuations roughly vary as k^{-2}, while the above non-equilibrium fluctuations vary as k^{-4}. Hence, as shown by Kirkpatrick, Ortiz de Zárate, and Sengers [Kirkpatrick *et al.* (2013, 2014)], Casimir effects in confined non-equilibrium fluids are substantially larger than critical Casimir effects in equilibrium fluids. As an example, we consider a liquid between two horizontal thermally conducting plates separated by a distance L and subject to a stationary temperature gradient ∇T. The non-equilibrium Casimir effects are two-fold. First, there will be a fluctuation-induced non-equilibrium contribution to the density profile as a function of height. Second, the fluctuations cause an additional non-equilibrium pressure contribution, $\bar{p}_{NE}$, to the equilibrium pressure such that [Kirkpatrick *et al.* (2013, 2014)]

$$\bar{p}_{NE} = \frac{c_p k_{\mathrm{B}} \overline{T_0}^2 (\gamma - 1)}{96\pi D_T (\nu + D_T)} \widetilde{B} F_0 L \left(\frac{\nabla T_0}{\overline{T_0}} \right)^2, \tag{3.16}$$

with,

$$\widetilde{B} = \left[1 - \frac{1}{\alpha c_p} \left(\frac{\partial c_p}{\partial T} \right)_p + \frac{1}{\alpha^2} \left(\frac{\partial \alpha}{\partial T} \right)_p \right], \tag{3.17}$$

Here, α is the thermal expansion coefficient, and γ is the ratio of the isobaric and isochoric heat capacities. The coefficient F_0 is a numerical constant whose value depends on the boundary conditions for the velocity fluctuations that are coupled with the temperature fluctuations through the temperature gradient. For stress-free boundary conditions, $F_0 = 1$. Just as in Eq.(3.15), all thermophysical properties, including the temperature T, can, to a good approximation, be identified with their average values in the liquid layer. Note that, for a given value of the temperature gradient ∇T, the fluctuation-induced pressure increases with L. The physical reason is that the dependence of the fluctuations implies that in real space the correlations scale with the system size. This non-equilibrium pressure contribution corresponds to a nonlinear Onsager-like cross effect [Kirkpatrick *et al.* (2013, 2014)]:

$$\bar{p}_{NE} = \kappa_{NL}(\nabla T)^2, \tag{3.18}$$

where κ_{NL} is a coefficient in the Burnett equations mentioned earlier in Section 4. Comparing Eqs.(3.16) and (3.18), we see that the non-equilibrium fluctuation-induced pressure is directly related to the divergence of the nonlinear Burnett coefficient κ_{NL} with increasing L. Experimentally, it may be more convenient to investigate the fluctuation-induced pressure as a function of the temperature difference $\delta T = L\nabla T$. Then,

$$\bar{p}_{NE} \propto \frac{1}{L}\left(\frac{\delta T}{T}\right)^2 . \tag{3.19}$$

This result may be compared with the critical Casimir pressure in equilibrium fluids:

$$p_c = \frac{k_B T}{L^3}\Theta\left(\frac{L}{\xi}\right), \tag{3.20}$$

where ξ is the correlation length. From Eq. (3.20), we have estimated that for water at 298 K in a layer with 1 micron and with $\delta T = 25$ K, $\bar{p}_{NE}$ will be of the order of a Pa, while p_c is of the order of a milli-Pa for the same distance. Actually, at $L = 1$ mm, $\bar{p}_{NE}$ becomes already of the same order of magnitude as p_c at $L = 1$ micron. One should also note that the critical Casimir effect can only be observed in fluids near a critical point, while the non-equilibrium Casimir effect will be generically present in liquids at any temperature and density. We conclude that thermal fluctuations in non-equilibrium fluids are fundamentally different from thermal fluctuations in equilibrium fluids.

3.6 Long Time Tail Phenomena in Other Contexts

It is quite remarkable how often one encounters situations in other physical contexts where the long time tails or, equivalently, mode-coupling theory play an important role. Here we list just a few examples.

Critical Phenomena

As we mentioned earlier, mode-coupling theory was developed more or less intuitively by Kadanoff and Swift in order to explain the behavior of transport coefficients near the critical points of phase transitions, such as the liquid-gas transition. It was known from experiments of Sengers, carried out in the 1960's, that the coefficient of thermal conductivity diverges near this critical point, as recently reviewed by Anisimov [Anisimov (2011)]. In this situation, both static and dynamic correlations have a long range. Later experiments of Sengers and co-workers confirmed both effects [Chang *et al.* (1976); Burstyn and Chang (1980)]. The results were a combination of long time tail effects underlying mode-coupling theory with the effects of the singular behavior of thermodynamic properties of the fluid near its critical point. Other and related applications of mode-coupling theory to the liquid-glass transition have been important for the theory of glasses, but we will not comment on that work here.

Weak Localization

In Section 2, we mentioned that in the mid 1960$'s$ Langer and Neal [Langer and Neal (1966)] showed that a logarithmic term appears in the conductivity in disordered electron systems. It wasn't until the late 1970$'s$ that the dynamical consequences of the correlations that lead to the logarithmic term, basically quantum long-time tail effects, were studied and understood [Gorkov *et al.* (1979); Abrahams *et al.* (1979)].[6] This opened up the field of what became known as weak local-

[6] A review article that stresses the generality of long time tail phenomena in the context of a variety of closely related phenomena, both classical and quantum, is given in Ref. [Belitz *et al.* (2005)].

ization in condensed matter physics which in turn is closely connected to the phenomenon of Anderson localization [Anderson (1958)]. Among other things, the ultimate conclusion was that the effects were so strong that at zero temperature a two-dimensional system is always an insulator[Wegner (1979); Abrahams *et al.* (1979); Schäfer and Wegner (1980); Wegner (1980)]. At finite temperature, there are logarithmic temperature non-analyticities that decrease the conductivity as T is lowered. In three dimensions there are weaker, but still important non-analyticities in both temperature and frequency. All of these effects have been measured in great detail. For reviews, see Refs. [Lee and Ramakrishnan (1985); Belitz and Kirkpatrick (1994)].

Cosmology

There appears to be a deep and interesting connection between the long time tail phenomena that we have been discussing here and the results of investigations of the dynamics of black hole horizons. The cosmology community has become aware of the results of non-equilibrium statistical mechanics, in particular, the existence of long time tails and their anomalous effects on the equations of fluid dynamics. We will not go into the details, but it is worth mentioning the titles of a few recent papers: "Hydrodynamic Long Time Tails from Anti de Sitter Space" by S. Caron-Huot and O. Saremi [Caron-Huot and Saremi (2010)], and "Hydrodynamic Fluctuations, Long Time Tails, and Supersymmetry" by P. Kovtun and L. G. Yaffe [Kovtun and Yaffe (2003)], among others. Such connections reinforce the notion gained from experience that across a wide swath of physics, people, perhaps without being aware of it, are working on the same or closely related problems, and the only difference is in the mathematical language used to describe them.

3.7 Conclusion

The paper has given a brief review of the history of kinetic theory and related non-equilibrium statistical mechanics with an emphasis of the work of Alder and Wainwright as described in their 1970 paper. Alder and Wainwright helped consolidate prior work in kinetic theory and stimulated much more work in theoretical, experimental, and computational physics. We hope that we have made clear the profound influence the 1970 paper has had on non-equilibrium statistical mechanics and on fields that, on first sight, might seem to be distantly related, but on closer inspection, turn out to be closely related after all. We are pleased to dedicate this paper to our friend, colleague, and mentor, Berni Alder, on the occasion of his 90th birthday!

3.8 Acknowledgements

The authors would like to thank D. Belitz and D. Thirumalai for helpful discussions and Y. Bar Lev and A. Nava-Tudela for their considerable help with the preparation of this paper. They would also like to thank E. G. D. Cohen for useful and productive conversations over a period of many years. TRK would like to thank the NSF for support under Grant No. DMR-1401449

Bibliography

Abrahams, E., Anderson, P. W., Licciardello, D. C., and Ramakrishnan, T. V. (1979). *Phys. Rev. Lett* **42**, p. 673.

Adams, P. W., Browne, D., and Paalanen, M. A. (1992). *Phys. Rev. B* **45**, p. 8837.

Alder, B. J., Gass, D. M., and Wainwright, T. E. (1970). *J. Chem. Phys.* **53**, p. 3813.

Alder, B. J. and Wainwright, T. E. (1967). *Phys. Rev. Lett.* **18**, p. 988.

Alder, B. J. and Wainwright, T. E. (1970). *Phys. Rev. A* **1**, p. 18.

Anderson, P. W. (1958). *Phys. Rev.* **109**, p. 1492.

Anisimov, M. A. (2011). *Int. J. Thermophys.* **32**, p. 2001.

Belitz, D. and Kirkpatrick, T. R. (1994). *Rev. Mod. Phys.* **66**, p. 261.

Belitz, D., Kirkpatrick, T. R., and Vojta, T. (2005). *Rev. Mod. Phys.* **77**, p. 579.

Bogoliubov, N. N. (1949). reprinted in J. de Boer and G. E. Uhlenbeck (eds.), *Studies in Statistical Mechanics*, Vol. 1 (North-Holland, Amsterdam), pp. 1–118.

Brush, S. G. (1972). *Kinetic Theory*, Vol. 3 (Pergamon, New York).

Burstyn, H. C. and Chang, R. F. (1980). *Phys. Rev. Lett.* **44**, p. 410.

Caron-Huot, S. and Saremi, O. (2010). *J. High Ener. Phys.* **13**, p. 1.

Chang, R. F., Burstyn, H., Sengers, J. V., and Bray, A. J. (1976). *Phys. Rev. Lett.* **37**, p. 1481.

Choh, S. T. and Uhlenbeck, G. E. (1958). The kinetic theory of dense gases, Tech. rep., University of Michigan.

Cohen, E. G. D. (1962). *Physica* **28**, p. 1025.

Cohen, E. G. D. (1963). *J. Math. Phys.* **4**, p. 183.

Dorfman, J. R. (1981). *Physica A* **106**, p. 77.

Dorfman, J. R. and Cohen, E. G. D. (1965). *Phys. Lett.* **16**, p. 124.

Dorfman, J. R. and Cohen, E. G. D. (1967). *J. Math. Phys.* **8**, p. 282.

Dorfman, J. R. and Cohen, E. G. D. (1970). *Phys. Rev. Lett.* **25**, p. 1257.

Dorfman, J. R. and Cohen, E. G. D. (1972). *Phys. Rev. A* **6**, p. 776.

Dorfman, J. R. and Cohen, E. G. D. (1975). *Phys. Rev. A* **12**, p. 292.

Dorfman, J. R., Kirkpatrick, T. R., and Sengers, J. V. (1994). in *Ann. Rev. Phys. Chem.*, **45**, p. 213.

Ernst, M. H., Cichocki, B., Dorfman, J. R., Sharma, J., and van Beijeren, H. (1978). *J. Stat. Phys.* **18**, p. 237.
Ernst, M. H. and Dorfman, J. R. (1975). *J. Stat. Phys.* **12**, p. 311.
Ernst, M. H., Hauge, E. H., and van Leeuwen, J. M. J. (1970). *Phys. Rev. Lett.* **25**, p. 1254.
Erpenbeck, J. J. and Wood, W. W. (1991). *Phys. Rev. A* **43**, p. 4254.
Fisher, M. E. and de Gennes, P. (1978). *C. R. Acad. Sci. Paris B* **287**, p. 207.
Goldman, R. (1966). *Phys. Rev. Lett.* **17**, p. 130.
Gorkov, L. P., Larkin, A., and Khmelnitskii, D. E. (1979). *JETP Lett.* **30**, 228.
Green, M. S. (1952). *J. Chem. Phys.* **20**, p. 1281.
Green, M. S. (1954). *J. Chem. Phys.* **22**, p. 398.
Green, M. S. (1956). *J. Chem. Phys.* **25**, p. 836.
Green, M. S. (1964). *Phys. Rev.* **136**, p. 905.
Green, M. S. and Piccirelli, R. A. (1963). *Phys. Rev.* **132**, p. 1388.
Kadanoff, L. P. and Swift, J. (1966). *Phys. Rev.* **166**, p. 89.
Kamgar-Parsi, B. and Sengers, J. V. (1983). *Phys. Rev. Lett.* **51**, p. 2163.
Kardar, M. and Golestanian, R. (1999). *Rev. Mod. Phys.* **71**, p. 1233.
Kawasaki, K. (1970). *Ann. Phys.* **61**, p. 1.
Kawasaki, K. and Oppenheim, I. (1965). *Phys. Rev.* **139**, p. 1763.
Kirkpatrick, T. R. (1981). Ph.D. thesis, Rockefeller University, New York.
Kirkpatrick, T. R., Cohen, E. G. D., and Dorfman, J. R. (1982a). *Phys. Rev. A* **26**, p. 950.
Kirkpatrick, T. R., Cohen, E. G. D., and Dorfman, J. R. (1982b). *Phys. Rev. A* **26**, p. 972.
Kirkpatrick, T. R., Cohen, E. G. D., and Dorfman, J. R. (1982c). *Phys. Rev. A* **26**, p. 995.
Kirkpatrick, T. R., Ortiz de Zárate, J. M., and Sengers, J. V. (2013). *Phys. Rev. Lett.* **110**, p. 235902.
Kirkpatrick, T. R., Ortiz de Zárate, J. M., and Sengers, J. V. (2014). *Phys. Rev. E* **89**, p. 022145.
Kirkpatrick, T. R. and Dorfman, J. R. (2015). *Phys. Rev. E* **92**, p. 022109.
Kovtun, P. and Yaffe, L. G. (2003). *Phys. Rev. D* **68**, p. 025007.
Krech, M. (1994). *The Casimir Effect in Critical Systems* (World Scientific, Singapore).
Kubo, R. (1957). *J. Phys. Soc. Japan* **12**, p. 370.
Langer, J. S. and Neal, T. (1966). *Phys. Rev. Lett.* **16**, p. 984.
Lee, P. A. and Ramakrishnan, T. V. (1985). *Rev. Mod. Phys.* **57**, p. 287.
Morkel, C. and Gronemeyer, C. (1988). *Z. Phys. B* **72**, p. 433.
Onuki, A. (1979). *Phys. Lett. A* **70**, p. 31.
Pomeau, Y. (1971). *Phys. Rev. A* **3**, p. 1174.
Pomeau, Y. (1973). *Phys. Rev. A* **7**, p. 1134.
Pomeau, Y. and Resibois, P. (1975). *Phys. Rept.* **19**, p. 63.
Schäfer, L. and Wegner, F. (1980). *Z. Phys. B* **38**, p. 113.
Schwartz, K. (1980). *Phys. Rev. B* **21**, p. 5125.
Segrè, P. N., Gammon, R. W., Sengers, J. V., and Law, B. (1992). *Phys. Rev. A* **45**, p. 714.

Sengers, J. V. (1967). in W. E. Brittin (ed.), *Boulder Lectures in Theoretical Physics*, Vol. IX C (Gordon and Breach), pp. 335–374.

Wegner, F. (1979). *Z. Phys. B* **35**, p. 207.

Wegner, F. (1980). *Z. Phys. B* **36**, p. 209.

Wood, W. W. and Erpenbeck, J. J. (1976). *Ann. Rev. Phys. Chem.* **27**, p. 319.

Wysokinski, K. I., Park, W., Belitz, D., and Kirkpatrick, T. R. (1994). *Phys. Rev. Lett.* **73**, p. 2571.

Wysokinski, K. I., Park, W., Belitz, D., and Kirkpatrick, T. R. (1995). *Phys. Rev. E* **52**, p. 612.

Chapter 4

The Onset of Turbulence in Wall-bounded Flows with Surface Roughness and Fluctuations

Pratanu Roy, Todd H. Weisgraber, Berni J. Alder

Lawrence Livermore National Laboratory, Livermore, CA 94550 [1]

Abstract

The fundamental problem of turbulent transition in wall-bounded hydrodynamic flow is still not well understood. Previous studies of channel and pipe flows do not include the role of wall roughness and only consider linear treatment with fluctuations to unsuccessfully destabilize the flow. We investigate if small amplitude distributed wall roughness with and without fluctuations could initiate the transition to turbulence by direct numerical simulation using a lattice-Boltzmann method that is equivalent to solving the nonlinear Navier-Stokes equation. The results show how a single roughness feature with microscopic amplitude on an otherwise smooth wall can generate the transition to turbulence in plane Poiseuille flow. The effect of fluctuations with smooth walls or the combined effect of fluctuations and wall perturbances predicts the onset of turbulence at about the same critical Reynolds number which is about twice the experimental one.

4.1 Introduction

Turbulent fluid flow is often viewed as the biggest unsolved problem in classical physics. Understanding the transition from a predictable laminar state to a disordered turbulent flow has remained an open challenge since the pioneering experiments of Reynolds over a century ago [Reynolds

[1] Telephone: +1-925-423-5917, E-mail: roy23@llnl.gov

(1883)]. One of the key issues lies in reconciling the predictions from linear stability analysis with experimental observations. In stability analysis, the Navier-Stokes equations governing hydrodynamics are linearized to identify eigenmodes which grow exponentially in time [Drazin and Reid (1981)]. Stability criteria are established based on independent physical parameters, often expressed in a single dimensionless variable, of which the most common one is the Reynolds number (Re), $Re = \frac{UD}{\nu}$. For wall-bounded flows, U is the average velocity, D is the channel height or pipe diameter, and ν is the kinematic viscosity of the fluid.

For channel flows bounded by planar walls, linear stability theory predicts transition at a critical Reynolds number, $\mathrm{Re}_c \approx 7,700$ [Orszag (1971)]. Conversely, flow in a circular pipe is stable to linear perturbations at all Re. However, typical experimental observations show transition at much smaller Re_c for both flows ($\approx 1,400$ for channels and $\approx 2,000$ for pipes). Recent experimental and computational studies involve the artificial introduction of a finite amplitude perturbation to the laminar flow field and the observation of the growth or decay of this disturbance. It is now recognized from these studies that two limitations of stability analysis are responsible for this discrepancy between theory and experiment: assumption of linearity, and contributions from non-eigenmode perturbations. In pipe flow, wall roughness could be the source of instabilities [Mullin (2011)], and in channel flows, non-linearities not treated by the theoretical analysis could significantly amplify and promote transition [Trefethen *et al.* (1993)].

One essential element that is missing from previous investigations, which assume completely smooth boundaries, is the role of wall roughness in generating the flow disturbances. Experiments clearly show the link between transition and roughness since carefully smoothed walls of pipes can increase the critical Reynolds number by orders of magnitude [Pfenninger (1961)]. However, the type of disturbances generated by roughness elements on natural surfaces and the relationship between them and the onset of transition remain unknown. Furthermore, hydrodynamic fluctuations, also known to affect flow stability [Kadau *et al.* (2007)], are neglected in analytic and computational models of transition [de Zárate and Sengers (2006, 2008)].

In spite of considerable experimental evidence that wall roughness is an important factor in the onset of turbulence, there have been few investigations, theoretically or computationally, of the effect of boundary conditions on flow stability. Part of the reason for the lack of such studies, is a fundamental belief among fluid dynamicists that such small protrusions on the wall could not possibly matter. It is, of course, difficult experimentally, computationally or theoretically to deal with such a large range of length scales, from the nearly microscopic to the macroscopic and it is therefore convenient to ignore the very small scale. There exist well known dimensional arguments based on the viscous length scale, δ_ν, that rationalize that below a certain amplitude, k, the wall roughness can be neglected. This region, where $k^+ = k/\delta_\nu < 5$, is known as the viscous sub-layer where the viscous stress dominates over the turbulent stress. Experiments have shown that for amplitudes below the sub-layer, the fully developed turbulent friction is independent of the roughness [Schlichting and Gersten (2000)], however the effect of these small amplitudes during transition are still unknown.

We conjecture that introducing realistic boundary conditions to represent roughness and incorporating hydrodynamic fluctuations are crucial to quantitatively understand the turbulent transition process observed in wall-bounded flows. Therefore, we study the physics of roughness-induced transition in a plane Poiseuille flow. We demonstrate the evolution of primary instability from a local perturbation arising from a single roughness amplitude to a global transition to turbulence. In addition, hydrodynamic fluctuations are incorporated into the equations to ascertain their influence on flow instability. It is very likely that the combination of fluctuation waves interacting with wall protrusions could resolve the onset of turbulence in wall-bounded flows. We feel that this possibility has been ignored in the literature and should certainly be explored.

4.2 Computational Methods

Our computer experiments use the lattice-Boltzmann method (LBM), which we showed some time ago to exactly reproduce the Navier-Stokes momentum conservation equations when 19 different velocities are used. We believe the energy conservation equation is not necessary to invoke since temperature changes are small in turbulence onset and would require 64 velocities to satisfy the full Navier-Stokes equations [McNamara *et al.* (1995)]. The LBM is preferable over traditional Navier-Stokes solvers for several reasons. Both have the essential requirement that adaptive mesh refinement (AMR) can be applied to span the length scales that have to be covered. With our approach to lattice-Boltzmann AMR, we have demonstrated the capability to reduce grid resolution by a factor of 32 using five levels of refinement [Guzik *et al.* (2014)]. This resolution is needed to resolve micron-scale wall roughness in a millimeter-sized channel. The advantage of using LBM over other Navier-Stokes solvers has to do with the ease of introducing boundary conditions. Since lattice-Boltzmann is basically a pseudo-particle method that evolves distribution functions for discrete velocities, the no-slip condition is enforced by simply reversing the outgoing distributions at the boundary. Another advantage of LBM is the ease with which fluctuations can be handled. When we tried to introduce fluctuations by the method suggested by Landau and Lifshitz (1987) into the Navier-Stokes equations, we found all existing Navier-Stokes solvers unstable. These solvers had been tested for stability at high wave number phenomena (shocks) and for general low wave number situations. However, at intermediate wave numbers, as needed for fluctuations which contain all wave numbers, they were found to be unstable unless one went to higher order schemes (third order Runge-Kutta) and even then they were still not completely stable [Donev *et al.* (2010)]. Lattice-Boltzmann experiences no such difficulties, one merely adds a fluctuating component to the 19 distributions. However, even the Boltzmann equation is unstable when it is discretized. We found it even more unstable than the corresponding Navier-Stokes equation. When discretizing Boltzmann into lattice-Boltzmann one loses the H-theorem which Boltzmann proved so beautifully to be satisfied, leading to unconditional numerical stability. One can, however, reintroduce an H-theorem in the LBM as a condition on each collision that guarantees numerical stability [Ansumali *et al.* (2003)]. We have shown that this works in the Taylor-Green vortex problem, where without the imposition of an H-theorem other numerical schemes are unstable. This is an enormous advantage when studying instabilities. With other Navier-Stokes solvers one is never sure whether a numerical instability occurs rather than a physical one.

4.3 Results and Discussion

Our preliminary computer experiments are direct simulations of transition to turbulence solely due to perturbations initiated by a single three-dimensional roughness element in plane Poiseuille flow. A computational domain of length L, width W, and height H were used, where $L = 2W = 3.5H$. The bump was placed in the middle of the bottom wall at ($L/2$, $W/2$, 0) and the bump size is 1% of the channel height. No-slip boundary conditions were used at the top and bottom walls and periodic boundary conditions were used along the streamwise and cross-stream direction. A laminar parabolic velocity profile was specified at the inlet as an initial condition of the simulation. Figure 4.1 shows a schematic of the computational domain with a single bump and initial inlet velocity profile.

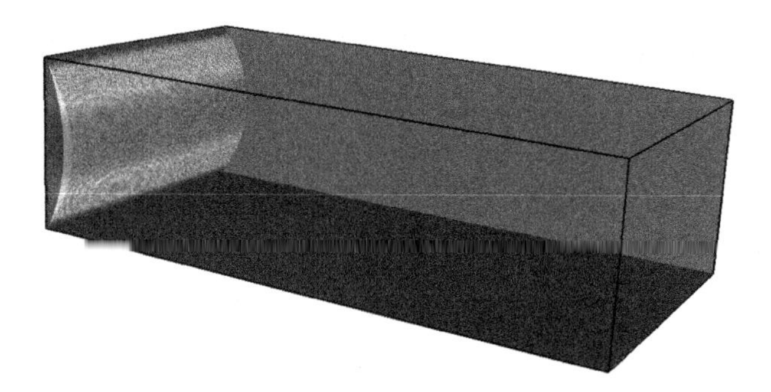

Fig. 4.1: Schematic of single bump (green) centered on the lower wall with the laminar velocity profile at the inlet

In order to study the onset of turbulence due to the effect of the bump, we gradually increased the Reynolds number starting from 2,000 up to 6,000. No external perturbations were added in these simulations. Figure 4.2 shows the snapshots of vorticity magnitude at the cross-section normal to the streamwise flow right above the bump at statistically steady-state conditions, when the average kinetic energy of the flow is constant. For low Reynolds number (Re = 2,000), the fully laminar flow was sustained throughout the duration of the simulation. With a slight increase in Reynolds number (Re = 2,400), steady oscillations with a wavelength of $W/2$ in the transverse direction were observed in the channel. However, these oscillations were stable without a sufficient amplitude to evolve to a turbulent state. As the Reynolds number was increased to 3,000, the wavelength of the secondary flow decreased to $W/3$ with an accompanying increase in amplitude. Again, the amplitude of these stable oscillations did not span the full channel height and was not sufficient to convert the flow into turbulence. Once the Reynolds number was increased to 3,600, unstable multi-mode oscillations were observed in the flow (not shown here), which eventually became three-dimensional leading to a fully turbulent flow. Subsequent increases in Reynolds number to Re = 4,500 and 6,000 exhibited quicker transitions to a turbulent state.

A more quantitative classification of transition can be achieved if we track the transverse kinetic energy with time, as shown in figure 4.3. Here, the transverse kinetic energy is defined as $KE_{avg} = \frac{1}{2} < v^2 + w^2 >$, where $<>$ denotes an averaging operator, v is the cross-stream velocity and w is the wall-normal velocity. In this case, the averaging was performed across the cross-stream (x-y) plane at different z-locations along the height of the channel. From this figure, we can identify two critical transitions in the flow: i) the onset of instability and ii) the onset of turbulence. The instability begins near the bump with oscillating modes, which gradually increases with time. Once the oscillations are fully formed with an amplitude spanning over the entire channel height, there is a sudden jump in transverse kinetic energy, which characterizes the onset of turbulence. At this point, the two-dimensional oscillating flow breaks down into a three-dimensional fully turbulent flow. Thus, a global transition of turbulence from a local finite amplitude perturbation arising from a single bump is evident in this picture.

In order to further characterize the transitional and turbulent flow, mean velocity profiles are plotted in figure 4.4. The time-averaged velocity profile was obtained once the flow reached a steady state as previously defined. The velocities were averaged in time along the channel height between the middle of the two periodic bumps, i.e. at the $(0, W/2, z)$ location. Since the lowest Reynolds number case (Re = 2,000) is not affected by the bump, the flow remains laminar and the velocity profile matches perfectly with the parabolic velocity profile of laminar flow. As the Reynolds number is increased to Re = 2,400, the velocity profile shows slight deviation from the standard laminar profile. The shifting of the maximum of mean velocity towards the bottom wall is due to the oscillations triggered by the bump. This effect is more evident for the Re = 3,000 case, where an increase in frequency and amplitude of the oscillations are observed. The maximum velocity is still shifted towards the bottom wall showing the transitional nature of the flow. It should be noted that, the shifting of the maximum velocity towards the bottom or top wall depends on the averaging location in the cross-stream direction. When the Reynolds number is increased to 3,600, the wall bias disappears and the velocity becomes more uniform near the centerline. This is due to the momentum redistribution caused by turbulence. Further increases in Reynolds number (Re = 4,500 and 6,000) do not result in any significant change in the mean velocities. We did not observe any relaminarization during these simulations but significantly longer run times would be required to confirm.

The mean velocity profiles are in good agreement with the experimental results reported by Lemoult *et al.* (2012), where they studied the subcritical transition to turbulence in plane Poiseuille flow. In order to perturb the flow, they used continuous injection from a water jet perpendicular to flow direction at the top wall. Instead of varying the Reynolds number, they varied the jet velocity for a fixed Reynolds number of about 3,800 [2]. The ratio u_r of maximum velocity to the initial center-line velocity was taken as a state parameter to define the onset of turbulence. For a fully turbulent flow, they reported a value of $u_r \approx 0.8$, which is similar to what we have found (0.77-0.79 for Re = 3,000-6,000). These slight discrepancies may be due to the difference in averaging procedure of the two studies. In addition to the time-averaging, they also conducted space-averaging to calculate the mean velocity profile. Nevertheless, the agreement with the experimental results gives us some level of confidence in validating our simulations.

To assess the computational ability to predict the critical Reynolds number, the skin friction coefficient (C_f) against Reynolds number is plotted in figure 4.5. Our simulation results are compared with the friction correlations for laminar and turbulent flow. The experimental results of Patel and Head (1969) are also plotted. The comparison shows that, the onset of transition occurs much earlier (around Re = 1,380) in the experiments, whereas in our simulations, we get laminar flow until Re = 2,000. Not only does the onset of transition occur at a later stage (Re = 2,400), the transition is also more gradual than the experiments. The fully turbulent friction coefficients match well with the Blasius friction correlation, but they underpredict Dean's correlation [Dean (1978)], which are in good agreement with the experimental data [Patel and Head (1969)]. It should be noted that, in almost all the experiments, artificial disturbances were used to generate turbulence, which might not be precisely controllable. The range of external disturbances at the entry region include edge effects, transverse injection from the channel walls, introduction of dye, and two-dimensional waves produced by a vibrating ribbon. In Patel and Head's experiments, the initiation of turbulence was attributed to the perturbations coming from the channel edges. Without any kind of perturbations, the laminar flow in the experiments was sustained up to Re = 5,500-8,000 using a regular rough walled surface. Therefore, it is very likely that the perturbation generated by a single bump with a size equal to 1% of the channel height is much lower than the perturbation amplitude generated by the external disturbances in

[2]The Re reported in the original paper was 2,830, which is based on the half-height and center-line velocity.

the experiments. This could possibly explain the discrepancies in the onset of transition between the experiments and our simulations.

We also investigated the effect of hydrodynamic fluctuations on the stability of the parallel plate channel flow. For each of the cases discussed above, fluctuating velocity components were calculated and added to the primary velocity components. We found that the hydrodynamic fluctuations introduced in the flow, along with a single bump, do not alter the onset of instability. However, they do affect the onset of turbulence. Figure 4.6 shows how the z-velocity varies with time (in lattice units, lu). The instability starts around $t = 4 \times 10^5$ lu and coincides with the case without fluctuations. However, the onset of turbulence occurs earlier with fluctuations. Our preliminary observation regarding this phenomena is that once the oscillation starts, the fluctuations add to the perturbations in cross-stream and wall-normal directions, and increases the transverse kinetic energy, thereby resulting in a quicker transition to turbulence.

4.4 Conclusion

This work is our yet incomplete attempt to find the mechanism for the onset of turbulence in wall-bounded flows with roughness elements and hydrodynamic fluctuations. We have demonstrated how a fully laminar flow undergoes transition through an oscillating flow and eventually transforms into a fully turbulent flow with the increase in Reynolds numbers. Although the onset of instability happens gradually, the breakdown of this two-dimensional oscillating flow into a three-dimensional turbulent flow occurs almost instantaneously throughout the channel. A quicker transition to turbulence is observed when fluctuations are added to the flow, although it did not alter the critical Reynolds number for the onset of instability or turbulence. Furthermore, fluctuations alone can initiate turbulence without any external perturbations. We believe that to get quantitative agreement with experiment, non-linear stress gradients must be included in the flow equations.

4.5 Acknowledgement

This work was performed under the auspices of the U.S. Department of Energy by Lawrence Livermore National Laboratory under Contract DE-AC52-07NA27344. The authors would like to thank Stephen Guzik and Phillip Colella for their contribution in developing the lattice-Boltzmann AMR code.

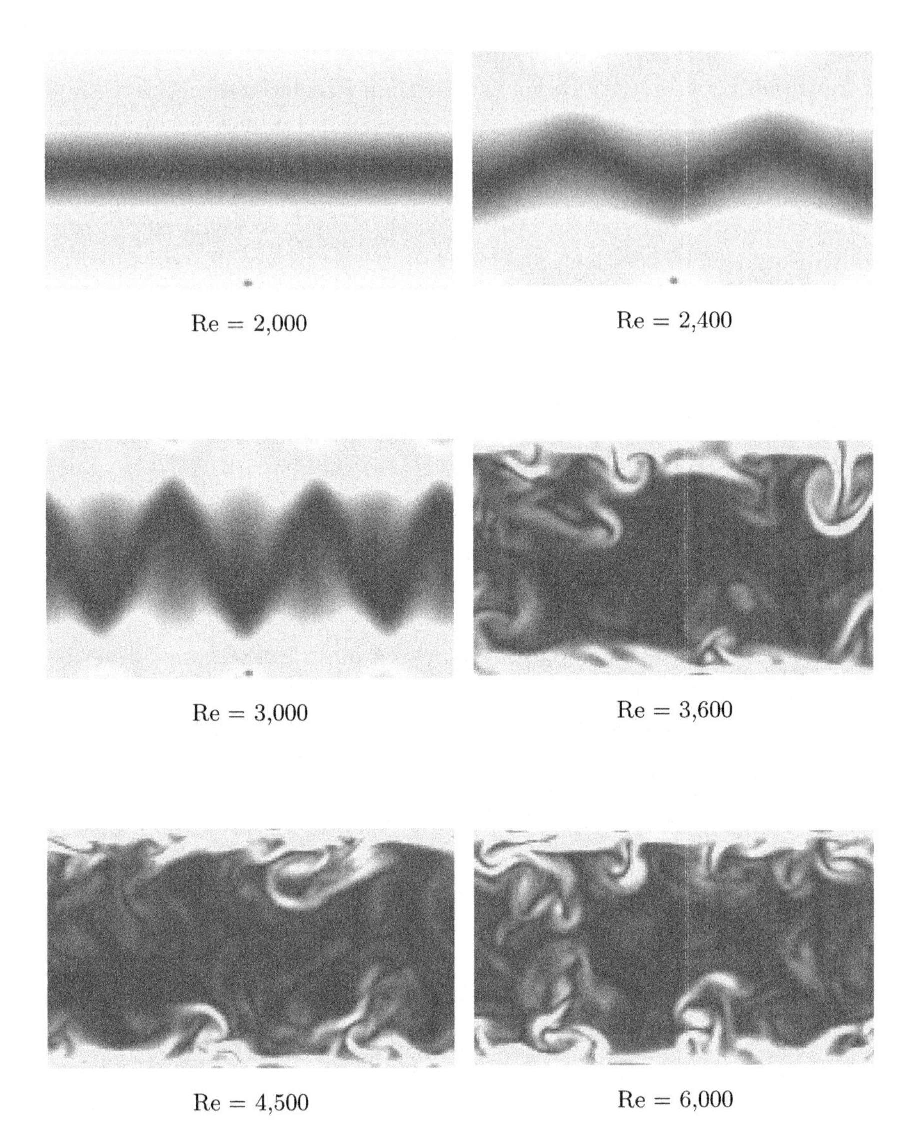

Fig. 4.2: Steady-state vorticity magnitude contours at different Reynolds numbers. The cross-sectional plane is taken at the location of the bump perpendicular to the primary flow direction. The blue end of the color spectrum represents the minimum vorticity magnitude, whereas the red end of the spectrum represents the maximum.

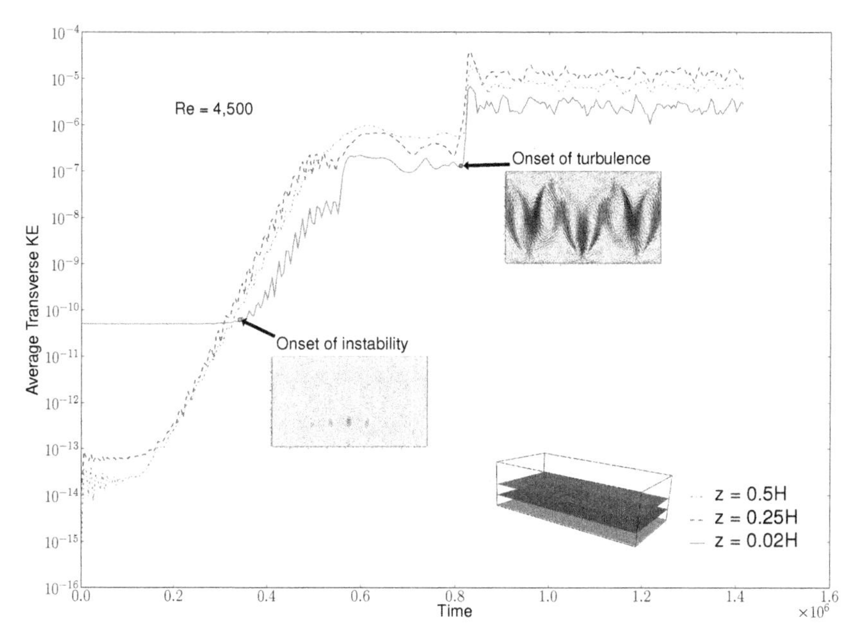

Fig. 4.3: The average cross-stream kinetic energy showing the onset of instability and turbulence for Re = 4,500. The wall-normal velocity contours are presented in the insets to illustrate the transitional flow structures.

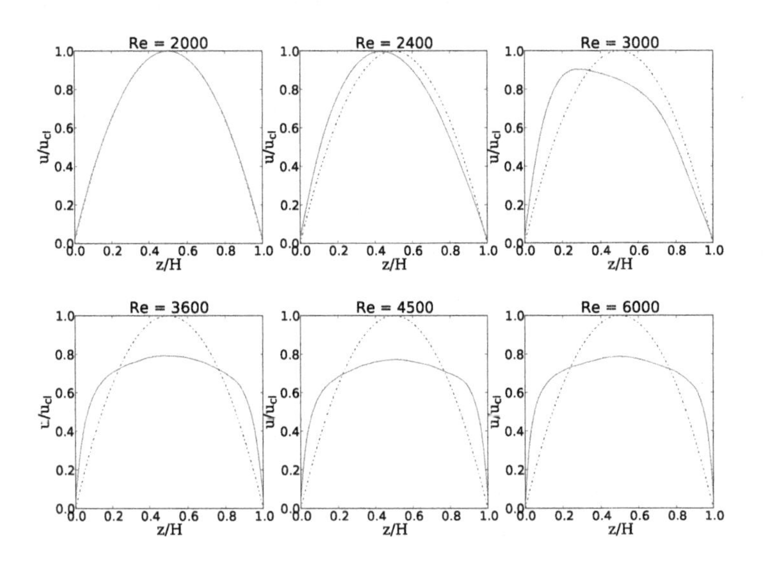

Fig. 4.4: Evolution of mean velocity profiles from fully laminar to fully turbulent states. The analytical laminar velocity profile is shown in a blue dash-dot line for comparison.

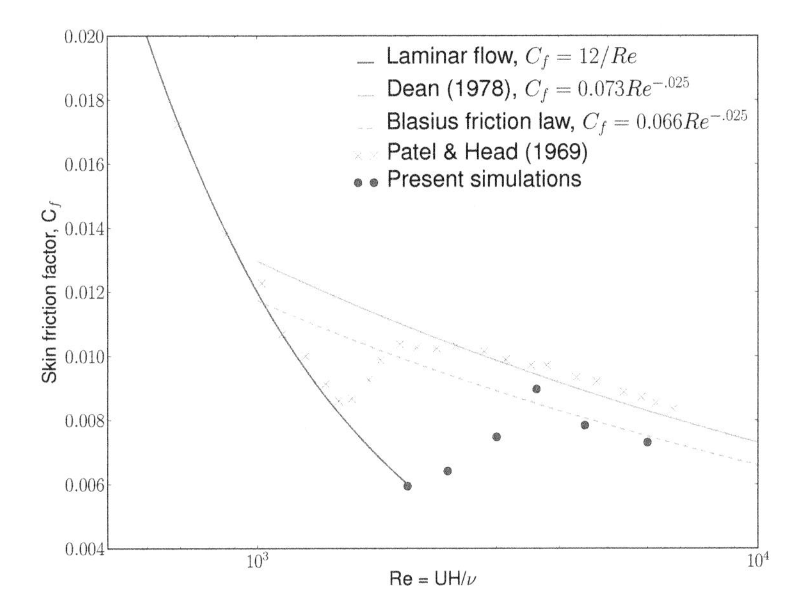

Fig. 4.5: Skin friction coefficient with Reynolds number. The solid cyan line represents Dean's correlation [Dean (1978)], and the dashed line represents the Blasius correlation. The cross markers represent the experimental results from Patel and Head (1969).

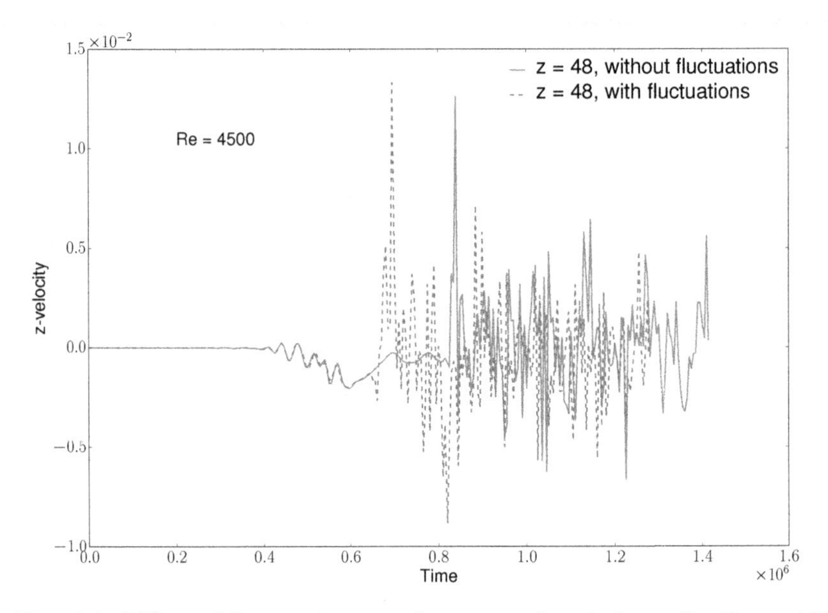

Fig. 4.6: Effect of fluctuations on the onset of turbulence for Re = 4,500. The z-velocity was computed at the center of the channel above the bump $(L/2, W/2, H/2)$. The velocity and time are shown in lattice units.

Bibliography

Ansumali, S., Karlin, I. V., and Ottinger, H. C. (2003). Minimal entropic kinetic models for hydrodynamics, *Europhys. Lett.* **63**, 6, pp. 798–804.

de Zárate, J. M. O. and Sengers, J. V. (2006). *Hydrodynamic fluctuations in fluids and fluid mixtures* (Elsevier).

de Zárate, J. M. O. and Sengers, J. V. (2008). Transverse-velocity fluctuations in a liquid under steady shear, *Physical Review E* **77**, 2, p. 026306.

Dean, R. (1978). Reynolds number dependence of skin friction and other bulk flow variables in two-dimensional rectangular duct flow, *Journal of Fluids Engineering* **100**, 2, pp. 215–223.

Donev, A., Vanden-Eijnden, E., Garcia, A., and Bell, J. (2010). On the accuracy of finite-volume schemes for fluctuating hydrodynamics, *Comm. App. Math Comp. Sci.* **5**, 2, pp. 149–197, doi:10.2140/camcos.2010.5.149.

Drazin, P. G. and Reid, W. H. (1981). *Hydrodynamic Stability* (Cambridge University Press).

Guzik, S. M., Weisgraber, T. H., Colella, P., and Alder, B. J. (2014). Interpolation methods and the accuracy of lattice-Boltzmann mesh refinement, *J. Comput. Phys.* **259**, C, pp. 461–487.

Kadau, K., Rosenblatt, C., Barber, J. L., Germann, T. C., Huang, Z., Carles, P., and Alder, B. J. (2007). The importance of fluctuations in fluid mixing, *Proc. Nat. Acad. Sci. USA* **104**, 19, pp. 7741–7745, doi: 10.1073/pnas.0702871104.

Landau, L. D. and Lifshitz, E. M. (1987). *Fluid mechanics*, Vol. 6 (Elsevier).

Lemoult, G., Aider, J.-L., and Wesfreid, J. E. (2012). Experimental scaling law for the subcritical transition to turbulence in plane poiseuille flow, *Physical Review E* **85**, 2, p. 025303.

McNamara, G. R., Garcia, A. L., and Alder, B. J. (1995). Stabilization of thermal lattice boltzmann models, *J. Stat. Phys.* **81**, 1-2, pp. 395–408, doi: 10.1007/BF02179986.

Mullin, T. (2011). Experimental studies of transition to turbulence in a pipe, **43**, pp. 1–24.

Orszag, S. (1971). Accurate solution of the Orr-Sommerfeld stability equation, *J. Fluid Mech.* **50**, pp. 689–703, doi:10.1017/S0022112071002842.

Patel, V. and Head, M. (1969). Some observations on skin friction and velocity profiles in fully developed pipe and channel flows, *Journal of Fluid Mechanics* **38**, 01, pp. 181–201.

Pfenninger, W. (1961). Transition in the inlet length of tubes at high reynolds numbers, in G. Lachman (ed.), *Boundary layer and flow control* (Pergamon), pp. 970–980.

Reynolds, O. (1883). An experimental investigation of the circumstances which determine whether the motion of water shall be direct or sinuous, and of the law of resistance in parallel channels. *Proceedings of the royal society of London* **35**, 224-226, pp. 84–99.

Schlichting, H. and Gersten, K. (2000). *Boundary-Layer Theory*, 8th edn. (Springer-Verlag).

Trefethen, L., Trefethen, A., Reddy, S., and Driscoll, T. (1993). Hydrodynamic stability without eigenvalues, *Science* **261**, 5121, pp. 578–584, doi: 10.1126/science.261.5121.578.

Chapter 5

Reversible Diffusion by Thermal Fluctuations

Aleksandar Donev, Thomas G. Fai and Eric Vanden-Eijnden

Courant Institute of Mathematical Sciences
New York University
New York, NY 10012

Abstract

A model for diffusion in liquids that couples the dynamics of tracer particles to a fluctuating Stokes equation for the fluid is investigated in the limit of large Schmidt number. In this limit, the concentration of tracers is shown to satisfy a closed-form stochastic advection-diffusion equation that is used to investigate the collective diffusion of hydrodynamically-correlated tracers through a combination of Eulerian and Lagrangian numerical methods. This analysis indicates that transport in liquids is quite distinct from the traditional Fickian picture of diffusion. While the ensemble-averaged concentration follows Fick's law with a diffusion coefficient that obeys the Stokes-Einstein relation, each instance of the diffusive mixing process exhibits long-ranged giant fluctuations around its average behavior. We construct a class of mesoscopic models for diffusion in liquids at different observation scales in which the renormalized diffusion coefficient depends on this scale. This indicates that the Fickian diffusion coefficient in liquids is not a material constant, but rather, changes with the scale at which experimental measurements are performed.

This summary article is devoted to Berni Alder on the occasion of his 90th birthday, both for his pioneering work on the importance of hydrodynamics to diffusion in liquids, and for being a great mentor to many scientists over the decades. A. Donev would in particular like to thank Berni for introducing him to the field of fluctuating hydrodynamics, for being a mentor and friend, and above all, for being an inspiration.

Diffusion is one of the most ubiquitous transport process. It is, arguably, the simplest dissipative mechanism. Fick's law of diffusion is "derived" in most elementary textbooks, and relates diffusive fluxes to the gradient of chemical potentials via a diffusion coefficient that is typically thought of as a material property. Yet, there are several hints that diffusion in liquids is, in fact, rather subtle. A first hint is that the Stokes-Einstein (SE) prediction for the diffusion coefficient involves the viscosity of the fluid, a seemingly independent transport property. This suggests a connection between momentum transport and diffusion and may explain why the SE prediction is in surprisingly reasonable agreement with measurements even in cases where it should not apply at all, such as molecular diffusion. A second hint is that nonequilibrium diffusive transport is accompanied by "giant" long-range correlated thermal fluctuations [Dorfman *et al.* (1994); Brogioli *et al.* (2000); Zarate and Sengers (2006)], which have been measured using light scattering and shadowgraphy techniques [Vailati and Giglio (1997); Brogioli *et al.* (2000); Croccolo *et al.* (2007); Vailati *et al.* (2011)]. Over the past fifty years, there has been an extensive effort to understand the diffusion of tracer particles in liquids, stemming in large part from the seminal discovery of Alder and Wainwright that hydrodynamics leads to a long-time tail in the velocity autocorrelation function of liquid molecules [Alder and Wainwright (1970)]. It is now well-understood that these unexpected features of diffusion in liquids stem from the contributions of advection by thermal velocity fluctuations [Bedeaux and Mazur (1974); Brogioli and Vailati (2000); Donev *et al.* (2011b); Detcheverry and Bocquet (2012)]. It has long been appreciated in statistical mechanics and nonequilibrium thermodynamics circles that thermal fluctuations exhibit long-ranged correlations in nonequilibrium settings [Dorfman *et al.* (1994); Zarate and Sengers (2006)]. The aim of this Letter is to show that these fluctuations are also of overarching importance to transport in fluids, a fact that has not been widely recognized so far.

In either gases, liquids or solids, one can, at least in principle, coarse-grain Hamiltonian dynamics for the atoms (at the classical level) to obtain a model of diffusive mass transport at hydrodynamic scales. This procedure is greatly simplified by first coarse-graining the microscopic dynamics to a simpler stochastic description, which is done by using kinetic theory for gases or Markov jump models for diffusion in solids. In both cases the picture that emerges is that of independent Brownian walkers performing uncorrelated random walks in continuum (gases) or on a lattice (solids). By contrast, in liquids the physical picture is rather different and must account for hydrodynamic correlations among the diffusing particles. In a liquid, molecules become trapped (caged) over long periods of time, as they collide with their neighbors. Therefore, momentum and energy are exchanged (diffuse) much faster than the molecules themselves can escape their cage. The main mechanism by which molecules diffuse is the motion of the whole cage when a large-scale velocity fluctuation (coordinated motion of parcels of fluid) moves a group of molecules and shifts and rearranges the cage.

Most previous theoretical studies of molecular diffusion are based on some form of mode-mode coupling, which is essentially a perturbative analysis in the strength of the thermal fluctuations [Bedeaux and Mazur (1974); Mazur and Bedeaux (1974); Bedeaux and Mazur (1975); Brogioli and Vailati (2000); Detcheverry and Bocquet (2012); Donev *et al.* (2011b)]. In this Letter we formulate a simple model for diffusion in liquids at microscopic and mesoscopic scales and use it to make a precise assessment of the contribution of fluctuations to diffusive transport. Our model is a simplified (coarse-grained) representation of the complex molecular processes that underlie mass transport in liquids. It mimics all of the crucial features of realistic liquids, while also being tractable analytically and numerically. Through a mix of theoretical and numerical studies, we show that this model exhibits realistic physical behaviors that differ from those of standard models of Fickian diffusion via uncorrelated random walks. In particular, we find that there is an unexpected connection between flows at small and large scales, and at microscopic and mesoscopic scales diffusion in liquids resembles turbulent diffusion.

Our model describes the motion of passive tracer particles advected by thermal velocity fluctuations, and can be used to describe the dynamics of fluorescently-labeled molecules in a Fluorescence Recovery After Photobleaching (FRAP) experiment, the transport of nano-colloidal particles in a nanofluid, or the motion of the molecules in a simple fluid. We will neglect direct interactions among the particles, which is appropriate when tracers are dilute. The evolution of the incompressible fluid velocity, $\boldsymbol{v}(\boldsymbol{r},t)$ with $\boldsymbol{\nabla}\cdot\boldsymbol{v}=0$, is assumed to satisfy the linearized fluctuating Navier-Stokes equation

$$\rho\partial_t\boldsymbol{v}+\boldsymbol{\nabla}\pi=\eta\boldsymbol{\nabla}^2\boldsymbol{v}+\sqrt{2\eta k_BT}\,\boldsymbol{\nabla}\cdot\boldsymbol{\mathcal{W}}, \tag{5.1}$$

where $\boldsymbol{\mathcal{W}}(\boldsymbol{r},t)$ denotes a white-noise symmetric tensor field (stochastic momentum flux) with covariance chosen to obey a fluctuation-dissipation principle [Zarate and Sengers (2006)],

$$\langle \mathcal{W}_{ij}(\boldsymbol{r},t)\mathcal{W}_{kl}(\boldsymbol{r}',t')\rangle = (\delta_{ik}\delta_{jl} + \delta_{il}\delta_{jk})\,\delta(t-t')\delta(\boldsymbol{r}-\boldsymbol{r}'). \tag{5.2}$$

The details of the microscopic coupling between the fluid and the passive tracer are complicated [Hynes *et al.* (1979)] and some approximations are required. We will assume that the evolution of the concentration of a large collection of tracers, $c(\boldsymbol{r},t)$, can be modeled via a fluctuating advection-diffusion equation

$$\partial_t c = -\boldsymbol{u}\cdot\boldsymbol{\nabla}c + \chi_0\boldsymbol{\nabla}^2 c. \tag{5.3}$$

where χ_0 is the *bare* (molecular) diffusion coefficient and the advecting velocity $\boldsymbol{u}$ is obtained by convolving the fluid velocity $\boldsymbol{v}$ with a smoothing kernel $\boldsymbol{K}_\sigma$ that filters out features below the molecular scale σ,

$$\boldsymbol{u}(\boldsymbol{r},t) = \int \boldsymbol{K}_\sigma(\boldsymbol{r},\boldsymbol{r}')\,\boldsymbol{v}(\boldsymbol{r}',t)\,d\boldsymbol{r}' \equiv (\boldsymbol{K}_\sigma \star \boldsymbol{v})(\boldsymbol{r},t), \tag{5.4}$$

and preserves the zero-divergence condition, $\boldsymbol{\nabla}\cdot\boldsymbol{u} = 0$. Physically, one can think of σ as representing the size of the molecular cage in the case of molecular diffusion and the radius of the tracer particles for colloidal diffusion. We stress that the smoothing of the fluctuating velocity field $\boldsymbol{v}$ is necessary to avoid divergence (ultraviolet catastrophe) of the effective diffusion coefficient of the tracer particle obtained below. Thus, the molecular scale details enter (5.3) in two ways: through the term $\chi_0\boldsymbol{\nabla}^2 c$ *and* the smoothing of $\boldsymbol{u}$ below scale σ. As we will see below, the smoothing turns out to be more important for transport than the first. Molecular dynamics simulations have confirmed that (5.1) and (5.3) accurately model diffusive mixing between two initially phase-separated fluids down to essentially molecular scales [Donev *et al.* (2014b)].

Note that an additional (mathematically problematic) multiplicative noise term $\boldsymbol{\nabla}\cdot\left(\sqrt{2\chi_0 c}\,\boldsymbol{\mathcal{W}}_c\right)$, where $\boldsymbol{\mathcal{W}}_c(\boldsymbol{r},t)$ is a white noise random vector field, needs to be included in (5.3) to capture equilibrium concentration fluctuations [Donev *et al.* (2014a)]; we do not include this term in this work in order to focus our attention on the nonequilibrium (giant) fluctuations that appear due to the advection by the fluctuating velocity.

In liquids, diffusion of mass is much slower than that of momentum, i.e. the velocity evolves fast compared to the concentration. This separation of time scales is measured by the Schmidt number and it can be used to eliminate the velocity [Papanicolaou (1976); Majda *et al.* (2006)]. This procedure, the details of which are presented elsewhere [Donev *et al.*

(2014a)], gives a *limiting* stochastic advection-diffusion equation for the concentration which reads [1]

$$\begin{aligned} \partial_t c &= -\boldsymbol{w} \odot \boldsymbol{\nabla} c + \chi_0 \boldsymbol{\nabla}^2 c & \text{(S)} \\ &= -\boldsymbol{w} \cdot \boldsymbol{\nabla} c + \chi_0 \boldsymbol{\nabla}^2 c + \boldsymbol{\nabla} \cdot [\boldsymbol{\chi}(\boldsymbol{r}) \boldsymbol{\nabla} c] & \text{(I)} \end{aligned} \tag{5.5}$$

where the first equality shows the equation in Stratonovich's interpretation and the second in Ito's. Here the advection velocity $\boldsymbol{w}(\boldsymbol{r}, t)$ is divergence free ($\nabla \cdot \boldsymbol{w} = 0$) and white-in-time, with covariance proportional to a Green-Kubo integral of the auto-correlation function of $\boldsymbol{u}(\boldsymbol{r}, t)$, i.e. $\langle \boldsymbol{w}(\boldsymbol{r}, t) \otimes \boldsymbol{w}(\boldsymbol{r}', t') \rangle = \boldsymbol{\mathcal{R}}(\boldsymbol{r}, \boldsymbol{r}') \, \delta(t - t')$ where

$$\boldsymbol{\mathcal{R}}(\boldsymbol{r}, \boldsymbol{r}') = 2 \int_0^\infty \langle \boldsymbol{u}(\boldsymbol{r}, t) \otimes \boldsymbol{u}(\boldsymbol{r}', t + t') \rangle dt', \tag{5.6}$$

and the enhancement of the diffusion coefficient is $\boldsymbol{\chi}(\boldsymbol{r}) = \frac{1}{2} \boldsymbol{\mathcal{R}}(\boldsymbol{r}, \boldsymbol{r})$. Similar equations, but with a distinct form of the covariance $\boldsymbol{\mathcal{R}}$, appear in the Kraichnan model of turbulent transport [Eyink and Xin (2000); Chaves *et al.* (2001)] (see Sec. 4.1 in [Majda and Kramer (1999)]). It can be shown [Donev *et al.* (2014a)] that at the Lagrangian level (individual tracer trajectories) (5.5) is equivalent to the well-known equations of Brownian Dynamics with hydrodynamic interactions (correlations) of a form similar to the Rotne-Prager tensor [Delong *et al.* (2014)], which is widely used as a model for diffusion in dilute colloidal suspensions.

Equation (5.5) has properties that may seem paradoxical at first sight but have important implications for transport in liquids. Indeed notice that it is easy to take the average of this equation in Ito's form to deduce that the ensemble average of the concentration obeys Fick's law,

$$\partial_t \langle c \rangle = \boldsymbol{\nabla} \cdot (\boldsymbol{\chi}_{\text{eff}} \boldsymbol{\nabla} \langle c \rangle) \quad \text{where} \quad \boldsymbol{\chi}_{\text{eff}} = \chi_0 \boldsymbol{I} + \boldsymbol{\chi}, \tag{5.7}$$

which is a well-known result that can be justified rigorously (c.f. Eq. (255) in [Majda *et al.* (2006)]) and holds even in the absence of bare diffusion, $\chi_0 = 0$. This is surprising considering that (5.5) is time-reversible when $\chi_0 = 0$, as made clear by Stratonovich's form of this equation. Furthermore, the same equation (5.7) holds for *all* moments of c when $\chi_0 = 0$. This is no contradiction: the "dissipative" term $\boldsymbol{\nabla} \cdot [\boldsymbol{\chi}(\boldsymbol{r}) \boldsymbol{\nabla} c]$ and the stochastic forcing term $-\boldsymbol{w} \cdot \boldsymbol{\nabla} c$ are signatures of the same physical process, advection by thermal velocity fluctuations. Including the first term but omitting the second violates fluctuation-dissipation balance and *cannot* be justified. For

[1] Here $\boldsymbol{w} \odot \boldsymbol{\nabla} c$ and $\boldsymbol{w} \cdot \boldsymbol{\nabla} c$ are short-hand notations for $\sum_k (\boldsymbol{\phi}_k \cdot \boldsymbol{\nabla} c) \circ d\mathcal{B}_k/dt$ and $\sum_k (\boldsymbol{\phi}_k \cdot \boldsymbol{\nabla} c) \, d\mathcal{B}_k/dt$, respectively, where $\mathcal{B}_k(t)$ are independent Wiener processes and $\boldsymbol{\phi}_k$ are basis functions such that $\boldsymbol{\mathcal{R}}(\boldsymbol{r}, \boldsymbol{r}') = \sum_k \boldsymbol{\phi}_k(\boldsymbol{r}) \otimes \boldsymbol{\phi}_k(\boldsymbol{r}')$.

example, the stochastic terms in (5.5) need to be retained to obtain the giant fluctuations seen in a *particular instance* (realization) of the diffusive mixing process.

Next we estimate the diffusion enhancement, $\boldsymbol{\chi}(\boldsymbol{r}) = \frac{1}{2}\boldsymbol{\mathcal{R}}(\boldsymbol{r},\boldsymbol{r})$, to get an intuitive understanding of its role. Since $\boldsymbol{v}(\boldsymbol{r},t)$ solves the linearized fluctuating Navier-Stokes equation (5.1), it is not hard to show that

$$\int_0^\infty \langle \boldsymbol{v}(\boldsymbol{r},t) \otimes \boldsymbol{v}(\boldsymbol{r}',t+t')\rangle dt' = \frac{k_B T}{\eta} \boldsymbol{G}(\boldsymbol{r},\boldsymbol{r}'), \tag{5.8}$$

where $\boldsymbol{G}$ is the Green's function (Oseen tensor) for the steady Stokes equation with unit viscosity, $\boldsymbol{\nabla}\pi = \boldsymbol{\nabla}^2\boldsymbol{v}+\boldsymbol{f}$ subject to $\boldsymbol{\nabla}\cdot\boldsymbol{v} = 0$ and appropriate boundary conditions. Inserting this expression in (5.6) implies that

$$\boldsymbol{\mathcal{R}}(\boldsymbol{r},\boldsymbol{r}') = \left(\boldsymbol{K}_\sigma \star \boldsymbol{G} \star \boldsymbol{K}_\sigma^T\right)(\boldsymbol{r},\boldsymbol{r}') \tag{5.9}$$

To proceed, recall that for an infinite isotropic system, $\boldsymbol{G}(\boldsymbol{r},\boldsymbol{r}') \equiv \boldsymbol{G}(\boldsymbol{r}-\boldsymbol{r}')$ is the Oseen tensor, the Fourier tranform of which reads $\widehat{\boldsymbol{G}}_{\boldsymbol{k}} = k^{-2}\left(\boldsymbol{I} - k^{-2}\boldsymbol{k}\otimes\boldsymbol{k}\right)$. Let us employ an isotropic filtering kernel $\boldsymbol{K}_\sigma$ that cuts off the fluctuations in the advective velocity $\boldsymbol{w}$ at *both* large and small scales to account respectively for the finite extent of the system L and the filtering at the molecular scale σ, and assume that the Fourier transform of $\boldsymbol{\mathcal{R}}(\boldsymbol{r}-\boldsymbol{r}') \equiv \boldsymbol{\mathcal{R}}(\boldsymbol{r},\boldsymbol{r}')$ is

$$\hat{\boldsymbol{\mathcal{R}}}_{\boldsymbol{k}} = \frac{2k_B T}{\eta} \frac{k^2 L^4}{(1+k^4L^4)(1+k^2\sigma^2)} \left(\boldsymbol{I} - \frac{\boldsymbol{k}\otimes\boldsymbol{k}}{k^2}\right) \tag{5.10}$$

Converting (5.10) to real space gives an isotropic enhancement to the diffusion tensor $\boldsymbol{\chi} = \boldsymbol{\mathcal{R}}(0)/2 = (2\pi)^{-d}\int\left(\hat{\boldsymbol{\mathcal{R}}}_{\boldsymbol{k}}/2\right)d\boldsymbol{k} = \chi\boldsymbol{I}$. This Fourier integral is exactly the one that appears in the linearized steady-state (static) approximate renormalization theory when $\nu \gg \chi_0$ [Bedeaux and Mazur (1974); Brogioli and Vailati (2000); Donev *et al.* (2011b)]. Here we obtain the same result with a simple, general, and precise calculation that gives [2] for $L \gg \sigma$

$$\chi \sim \frac{k_B T}{\eta} \begin{cases} (4\pi)^{-1}\ln(L/\sigma) & \text{if } d=2 \\ (6\pi\sigma)^{-1} & \text{if } d=3. \end{cases} \tag{5.11}$$

In three dimensions (5.11) gives the Stokes-Einstein prediction $\chi \sim \chi_{SE} = k_B T/(6\pi\eta\sigma)$ for the diffusion coefficient of a slowly-diffusing no-slip rigid sphere of radius σ. In two dimensions, the effective diffusion coefficient

[2] Some of the coefficients in (5.11) depend on the exact form of the spectrum $\hat{\boldsymbol{\mathcal{R}}}_{\boldsymbol{k}}$ in (5.10).

grows logarithmically with system size, in agreement with the Einstein relation and the Stokes paradox for the mobility of a disk of radius σ. This system-size dependence of the effective diffusion coefficient has been verified using steady-state particle simulations [Donev *et al.* (2011a,b)]. Note also that (5.11) allows us to validate *a posteriori* the assumption of large separation of time scales between concentration and momentum diffusion. Specifically, the limiting equation (5.5) is a good approximation to (5.3) if the effective Schmidt number $\mathrm{Sc} = \nu/\chi_{\mathrm{eff}} = \nu/(\chi_0 + \chi) \gg 1$. This is indeed the case in practice for simple liquids and macromolecular solutions.

The measured diffusion coefficients in molecular liquids and macromolecular solutions closely match the Stokes-Einstein prediction. This suggests that in realistic fluids diffusive transport is dominated by advection by the velocity fluctuations, $\chi \gg \chi_0$. Since we know that each realization follows a strictly reversible dynamics when $\chi_0 = 0$, but that the evolution of the mean is dissipative even in this case since $\chi_{\mathrm{eff}} = \chi > 0$, it is important to understand the difference in the behavior of the ensemble mean of the diffusive mixing process, described by (5.7), and the behavior of an individual realization, described by (5.5).

To this end, we resort to numerical experiments using finite-volume [Usabiaga *et al.* (2012)] Eulerian methods [3], as well as Lagrangian tracers algorithms [4] that we have developed specifically for the purpose of simulating the limiting dynamics (5.5) at $\chi_0 > 0$ and $\chi_0 = 0$, respectively. Details of these multiscale numerical methods are given elsewhere [Donev *et al.* (2014a)]. Let us consider the temporal decay of a smooth single-mode initial perturbation $c(\boldsymbol{r}, 0) = \sin(2\pi x/L)\sin(2\pi y/L)$ in two dimensions. The ensemble *mean* $\langle c \rangle$ follows the simple diffusion equation (5.7), and therefore remains a single-mode field with an amplitude decaying as $\exp(-t/\tau)$, where $\tau = \left(2\chi_{\mathrm{eff}} k_0^2\right)^{-1}$ is a decay time and k_0 is the initially

[3] We checked that the results of these simulations compare well to those obtained by integrating the resolved dynamics (5.1, 5.3) with $S_c = \nu/\chi_{\mathrm{eff}} \sim 10^3 - 10^4$ [Usabiaga *et al.* (2012)]. The overdamped simulations can reach the same time scales in *much* less (by a factor of about S_c) computational effort than the direct numerical simulation because they bypass the need to resolve the fast velocity fluctuations.

[4] The Lagrangian description associated with (5.5) reads

$$d\boldsymbol{q} = \sum_k \boldsymbol{\phi}_k(\boldsymbol{q}) \circ d\mathcal{B}_k + \sqrt{2\chi_0}\, d\boldsymbol{\mathcal{B}}_{\boldsymbol{q}},$$

where a *single* realization of the random field $\sum_k \boldsymbol{\phi}_k \circ d\mathcal{B}_k$ advects *all* of the walkers. This induces correlations between their trajectories which crucially affect the physics of the collective diffusion of the tracers.

excited wavenumber. In the inset of Fig. 5.1, we show a single *instance* (realization) of the concentration at time $t \approx 2\tau$ when $\chi_0 = 0$. The figure reveals characteristic *giant* (long-ranged) fluctuations in particular realizations of the diffusive process, with the contour lines of the concentration becoming rougher as time progresses [5]. These enhanced nonequilibrium fluctuations stem from the development of a power-law spectrum as the mixing progresses, as predicted by linearized fluctuating hydrodynamics [Zarate and Sengers (2006)]. The evolution of the power spectrum during the diffusive decay is illustrated in Fig. 5.1.

The conserved quantity $\int \left(c^2/2\right) d\boldsymbol{r}$ injected via the initial perturbation away from equilibrium is effectively dissipated through a mechanism similar to the energy cascade observed in turbulent flows. Advection transfers power from the large length scales to the small length scales, *effectively* dissipating the power injected into the large scales via the initial condition. A straightforward calculation that is detailed elsewhere [Donev *et al.* (2014a)] shows that the total rate at which power is lost ("dissipated") from mode $\boldsymbol{k}_0$ is given by $\boldsymbol{k}_0 \cdot \boldsymbol{\chi} \cdot \boldsymbol{k}_0$. This is exactly the same rate of dissipation as one would get for ordinary diffusion with diffusion tensor $\boldsymbol{\chi}$. However, in simple diffusion all other modes would remain unexcited and there would be no giant fluctuations.

At late times of the diffusive decay, $t \sim \tau$, one expects that a self-similar state will be reached in which the shape of the spectrum of c does not change as it decays exponentially in time as $\exp\left(-t/\tau\right)$. This is indeed what we observe, and the shape of the decaying spectrum is shown in Fig. 5.1. Numerically we observe that most of the bare dissipation occurs at the largest wavenumbers. Note however that the shape of the spectrum at the large wavenumbers is strongly affected by discretization artifacts and the presence of (small) bare diffusion. These numerical grid artifacts can be eliminated by using the Lagrangian tracer algorithm, which leads to a similar power-law behavior [Donev *et al.* (2014a)].

In the literature, linearized fluctuating hydrodynamics is frequently used to obtain the steady-state spectrum of fluctuations [Zarate and Sengers (2006)]. In the limit of large Schmidt numbers, the standard heuristic approach leads to the additive-noise equation,

$$\partial_t \tilde{c} = -\boldsymbol{w} \cdot \boldsymbol{\nabla} \langle c \rangle + \boldsymbol{\nabla} \cdot \left[\chi_{\text{eff}} \boldsymbol{\nabla} \tilde{c} \right], \tag{5.12}$$

where $\langle c \rangle$ is the ensemble mean, which follows (5.7). Note that we have not accounted for *equilibrium* concentration fluctuations in (5.12)

[5] We have performed hard-disk molecular dynamics simulation of this mixing process and observed the same qualitative behavior seen in the inset of Fig. 5.1.

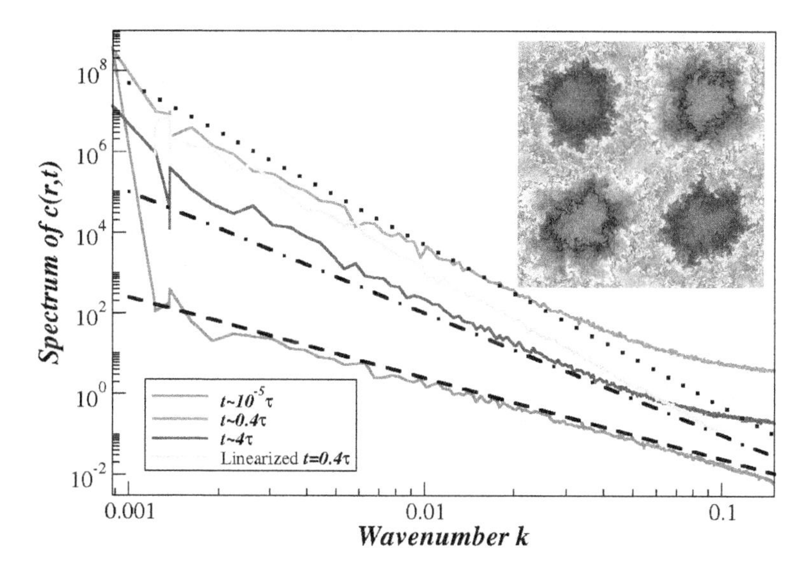

Fig. 5.1 The decay of a single-mode initial condition. The inset shows the concentration at a time $t \approx 2\tau$, as obtained from a Lagrangian simulation with 2048^2 tracers and no bare diffusion, along with numerical approximations to the contour lines. The main figure shows the power spectrum of an *individual* realization of the concentration $c(\boldsymbol{r}, t)$ at several times, as obtained using an Eulerian algorithm for solving (5.5). The power of individual modes $\boldsymbol{k}$ with nearby k is averaged and the result is shown with colored solid lines, while dashed/dotted lines show power laws k^{-2}, k^{-3} and k^{-4} for comparison. At early times $t \ll \tau = \left(2\chi_{\text{eff}} k_0^2\right)^{-1}$ (red line) power is being transferred from mode $\boldsymbol{k}_0 \approx 2\pi/L \approx 10^{-3}$, initially excited to have spectral power $p_{\boldsymbol{k}_0} \approx 7 \cdot 10^8$, to the rest of the modes, leading to a spectrum $\sim k^{-2}$. At late times $t \gtrsim \tau$ (magenta and blue lines), a steadily-decaying shape of the spectrum is reached where power transferred from the larger scales is dissipated at the small scales via bare diffusion. Numerically linearized fluctuating hydrodynamics predicts a spectrum $\sim k^{-4}$ (green line) [Zarate and Sengers (2006)].

since our focus here is on the *nonequilibrium* fluctuations and we wish to more accurately measure the power-law spectrum. Equation (5.12) can easily be solved analytically in the Fourier domain when $\boldsymbol{\nabla}\langle c\rangle = \boldsymbol{h}$ is a weak externally applied constant gradient to obtain a spectrum $(\boldsymbol{h} \cdot \hat{\boldsymbol{\chi}}_{\boldsymbol{k}} \cdot \boldsymbol{h}) / \left(\chi_{\text{eff}} k^2\right) \sim k^{-4}$ for intermediate wavenumbers. For finite gradients and more realistic boundary conditions, we can solve (5.12) numerically with the same algorithm used to solve the full nonlinear equation (5.5) by simply reducing the magnitude of the fluctuations by a large factor and then increasing the spectrum of the fluctuations by the same factor to obtain the spectrum of $\tilde{c}$ [Usabiaga *et al.* (2012)]. The result of this numerically-linearized calculation for the single-mode initial condition is shown in Fig. 5.1 and seen to follow the expected k^{-4} power-law [Zarate and Sengers (2006)]. This power law is not in a very good agreement with the spectrum obtained by solving the full nonlinear equation (5.5), which appears closer to k^{-3} in the two-dimensional setting we study here.

If there were only random advection, with no bare diffusion, the transfer of energy from the coarse to the fine scales would continue indefinitely, since the dynamics is reversible and there is nothing to dissipate the power. However, any features in c at length scales below molecular scales have no clear physical meaning. In fact, continuum models are inapplicable at those scales. In typical experiments, such as FRAP measurements of diffusion coefficients, one observes the concentration spatially-coarse grained at scales much larger than the molecular scale. It is expected that not resolving (coarse-graining) the microscopic scales will lead to true dissipation and irreversibility in the coarse-grained dynamics. Such coarse-graining can take form of ensemble averaging, or elimination of slow degrees of freedom. In either case, the loss of knowledge about the small scales will lead to positive entropy production.

It is reasonable to expect that one can replace the molecular scale details, or even all details of the dynamics at scales below some mesoscopic observation scale δ, by effective dissipation. In particular, we suggest that small-scale details in (5.5) can be replaced by a diffusive term with suitably chosen *renormalized* bare coefficient. This renormalization needs to be carried in such a way that the effective diffusion coefficient in the equation for the mean remains equal to $\boldsymbol{\chi}_{\text{eff}} = \chi_0 \boldsymbol{I} + \boldsymbol{\chi}$. A partial ensemble averaging of (5.5) can be used to achieve this goal [Donev *et al.* (2014a)], and this calculation leads to a spatially coarse-grained model for diffusion in liquids,

$$\partial_t c_\delta = -\boldsymbol{w}_\delta \odot \boldsymbol{\nabla} c_\delta + \boldsymbol{\nabla} \cdot \left[\left(\chi_0 \boldsymbol{I} + \Delta \boldsymbol{\chi}_\delta \right) \boldsymbol{\nabla} c_\delta \right], \tag{5.13}$$

where $c_\delta = \boldsymbol{K}_\delta \star c$ denotes the concentration filtered at the mesoscopic scale δ, the white-in-time random velocity $\boldsymbol{w}_\delta$ has covariance $\boldsymbol{K}_\delta \star \boldsymbol{\mathcal{R}} \star \boldsymbol{K}_\delta^T$, and χ_0 is renormalized by

$$\Delta \boldsymbol{\chi}_\delta(\boldsymbol{r}) = \frac{1}{2} \left(\boldsymbol{\mathcal{R}} - \boldsymbol{K}_\delta \star \boldsymbol{\mathcal{R}} \star \boldsymbol{K}_\delta^T \right) (\boldsymbol{r}, \boldsymbol{r}) \tag{5.14}$$

In Ito's form (5.13) is the same as (5.5) with $\boldsymbol{w}$ replaced by $\boldsymbol{w}_\delta$. Note that the renormalized *bare* diffusion coefficient $\boldsymbol{\chi}_0(\delta) = \chi_0 \boldsymbol{I} + \Delta \boldsymbol{\chi}_\delta$ in (5.13) is nonzero even if $\chi_0 = 0$. This true dissipation is a remnant of the unresolved (eliminated) small scales. However, it is important to stress that $\boldsymbol{\chi}_0(\delta)$ is not a material constant, but rather, depends on the mesoscopic lengthscale δ.

To test (5.13), consider diffusive mixing between two initially phase-separated fluids in two dimensions with periodic boundary conditions. We start with concentration $c = 1$ in a thin horizontal stripe, $c = 0$ everywhere. This could, for example, model a stripe in a FRAP experiment in which a laser beam combined with a diffraction grating is used to create a striped pattern of fluorescently labeled tracers at $t = 0$. In Fig. 5.2 we show snapshots of the concentration field at a later time, for $\delta = 0$ (no coarse-graining) in the top panel, and $\delta = 3\sigma$ in the middle and bottom panels. Specifically, in the middle panel we show the spatially smoothed concentration $\boldsymbol{K}_{\delta} \star c$. For comparison, in the bottom panel we show an instance of the solution of the proposed coarse-grained diffusion equation (5.13). Since $\boldsymbol{u}$ and $\boldsymbol{w}$ are spatially-smooth velocity fields, advection by these fields leads to behavior qualitatively different from diffusion when $\chi_0 = 0$. Specifically, if the initial concentration $c(\boldsymbol{r}, 0)$ has a sharp interface, this interface will remain sharp at all times, even if it becomes very rough, at *all* times, in *every* realization. Therefore, the top panel in the figure is black and white. In the presence of true (bare) dissipation, the interface between the two fluids does not remain sharp, and a range of concentrations $0 \leq c \leq 1$ appears for $t > 0$. Therefore, the middle and bottom panels in the figure show a spectrum of colors.

In large three dimensional systems, when the spatial coarse-graining is performed at macroscopic scales $\delta \gg \sigma$, it is expected that (5.13) will converge in some sense to the linearized fluctuating hydrodynamics equation (5.12), as suggested by renormalization arguments [Bedeaux and Mazur (1974)]. While we are not aware of mathematical tools to prove this type of statement, a plausible argument goes as follows. In three dimensions, as $\delta \to \infty$, the renormalization of the diffusion coefficient approaches the Stokes-Einstein value, $\Delta\boldsymbol{\chi}_{\delta} \to \boldsymbol{\chi}$, and the stochastic term $\boldsymbol{w}_{\delta} \odot \boldsymbol{\nabla} c_{\delta}$ becomes negligible because most of the power (spectral intensity) in the the random advection velocity $\boldsymbol{w}_{\delta}$ is removed by the filtering (since the spectrum of $\boldsymbol{w}$ decays like k^{-2}, in three dimensions the power distribution is independent of k). Therefore the noise will become "weak" in a suitable sense and the fluctuations can be linearized around the mean. This is not true in two dimensions, where large scale features in $\boldsymbol{w}_{\delta}$ give the dominant contribution to the effective diffusion and contain the majority of the spectral power of $\boldsymbol{w}$ (in two dimensions the power distribution decays like k^{-1}). Therefore, linearization is certainly not appropriate in two dimensions even if $\delta \gg \sigma$. Thin films may exhibit an intermediate behavior depending on the scale of observation relative to the thickness of the thin film [Bechhoefer *et al.* (1997)].

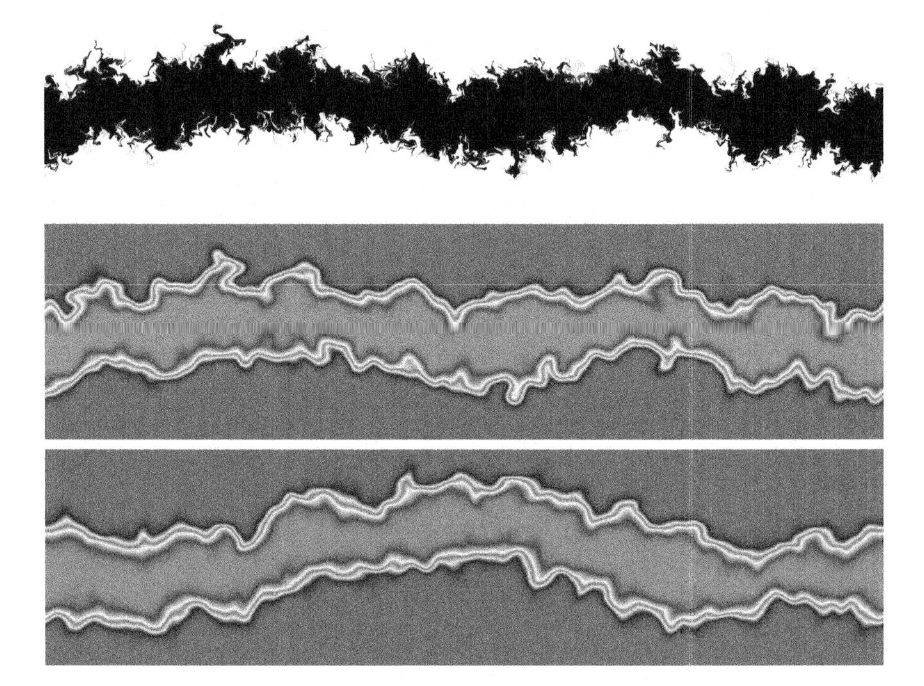

Fig. 5.2 *Top panel*: A snapshot of the concentration c for diffusive mixing of two miscible fluids in the absence of bare diffusion, starting from concentration being unity (black) in a horizontal stripe occupying one third of the periodic domain, and zero (white) elsewhere. The top and bottom interface are represented with about half a million Lagrangian tracers each. *Middle panel*: The spatially-coarse grained concentration c_δ obtained by blurring the top panel using a Gaussian filter with standard deviation $\delta = 3\sigma$. *Bottom panel*: An independent snapshot of the spatially coarse-grained concentration c_δ at the same point in time as the top panel, obtained by solving (5.13) with an Eulerian method using a grid of 2048×512 finite-volume cells. A Gaussian filter of width $\delta = 3\sigma$ is used to filter the discrete velocity. The effective diffusion coefficient χ_{eff} is the same as in the top panel.

Summing up, in both two and three dimensions the behavior of mixing processes in liquids cannot be described by Fick's law at mesoscopic scales. One must include random advection by the mesoscopic scales of the velocity fluctuations in order to reproduce not just the behavior of the mean but also the long-range correlated fluctuations observed in individual realizations. This emphasizes the crucial distinction between the self-diffusion of *individual* tracers and the *collective* diffusion of many hydrodynamically-correlated tracers. The traditional Fick's diffusion constant is only meaningful under special conditions (e.g., large three-dimensional bulk systems observed at macroscopic scales) which may not in fact be satisfied in experiments aimed to measure "the" diffusion coefficient. A length scale of observation (coarse-graining) must be attached to the diffusion coefficient value in order to make it a "material constant" that can be used in a predictive model of diffusive transport [Donev *et al.* (2011b)]. Furthermore, the measured diffusion coefficient is strongly affected by boundary conditions (confinement) [Detcheverry and Bocquet (2012); Bechhoefer *et al.* (1997); Donev *et al.* (2011a)].

We hope that these results will spur interest in designing experiments that carefully examine diffusion at a broad range of length scales. Existing experiments have been able to measure concentration fluctuations across a wide range of lenghtscales transverse to the gradient, but fluctuations are averaged longitudinally over essentially macroscopic scales (thickness of the sample) [Vailati and Giglio (1997); Croccolo *et al.* (2007); Vailati *et al.* (2011)]. While FRAP experiments routinely look at diffusion at micrometer scales, we are not aware of any work that has even attempted to account for the effect of thermal fluctuations. Giant fluctuations are expected to be more easily observed in thin liquid films due to the quasi-two dimensional geometry [Bechhoefer *et al.* (1997); Schulz *et al.* (2010)]. In the future we will consider extensions of our approach to multispecies liquid mixtures. Such extensions are expected to lead to a better understanding of the physics of diffusion in fluid mixtures, including a generalized Stokes-Einstein relation for inter-diffusion coefficients in dilute multispecies solutions.

Acknowledgments

We would like to acknowledge Florencio Balboa Usabiaga and Andreas Klockner for their help in developing a GPU implementation of the numerical methods, and Leslie Greengard for advice on the non-uniform FFT algorithm. We are grateful to Alberto Vailati, Alejandro Garcia, John Bell, Sascha Hilgenfeldt, Mike Cates and Ranojoy Adhikari for their insightful comments. A. D. was supported in part by the NSF grant DMS-1115341 and the DOE Early Career award DE-SC0008271. T. F.was supported in part by the DOE CSGF grant DE-FG02-97ER25308. E. V.-E. was supported by the DOE ASCR grant DE-FG02-88ER25053, the NSF grant DMS07-08140, and the ONR grant N00014-11-1-0345.

Bibliography

Alder, B. J. and Wainwright, T. E. (1970). Decay of the velocity autocorrelation function, *Phys. Rev. A* **1**, 1, pp. 18–21.

Bechhoefer, J., Géminard, J.-C., Bocquet, L., and Oswald, P. (1997). Experiments on tracer diffusion in thin free-standing liquid-crystal films, *Phys. Rev. Lett.* **79**, pp. 4922–4925.

Bedeaux, D. and Mazur, P. (1974). Renormalization of the diffusion coefficient in a fluctuating fluid I, *Physica* **73**, pp. 431–458.

Bedeaux, D. and Mazur, P. (1975). Renormalization of the diffusion coefficient in a fluctuating fluid III. Diffusion of a Brownian particle with finite size, *Physica A Statistical Mechanics and its Applications* **80**, pp. 189–202.

Brogioli, D. and Vailati, A. (2000). Diffusive mass transfer by nonequilibrium fluctuations: Fick's law revisited, *Phys. Rev. E* **63**, 1, p. 12105.

Brogioli, D., Vailati, A., and Giglio, M. (2000). Universal behavior of nonequilibrium fluctuations in free diffusion processes, *Phys. Rev. E* **61**, pp. R1–R4.

Chaves, M., Eyink, G., Frisch, U., and Vergassola, M. (2001). Universal decay of scalar turbulence, *Phys. Rev. Lett.* **86**, pp. 2305–2308.

Croccolo, F., Brogioli, D., Vailati, A., Giglio, M., and Cannell, D. S. (2007). Nondiffusive decay of gradient-driven fluctuations in a free-diffusion process, *Phys. Rev. E* **76**, 4, p. 041112.

Delong, S., Usabiaga, F. B., Delgado-Buscalioni, R., Griffith, B. E., and Donev, A. (2014). Brownian Dynamics without Green's Functions, *J. Chem. Phys.* **140**, 13, 134110, doi:10.1063/1.4869866, software available at `https://github.com/stochasticHydroTools/FIB`.

Detcheverry, F. and Bocquet, L. (2012). Thermal fluctuations in nanofluidic transport, *Phys. Rev. Lett.* **109**, p. 024501.

Donev, A., Fai, T. G., and Vanden-Eijnden, E. (2014a). A reversible mesoscopic model of diffusion in liquids: from giant fluctuations to Fick's law, *Journal of Statistical Mechanics: Theory and Experiment* **2014**, 4, p. P04004, `http://stacks.iop.org/1742-5468/2014/i=4/a=P04004`.

Donev, A., Garcia, A. L., de la Fuente, A., and Bell, J. B. (2011a). Diffusive Transport by Thermal Velocity Fluctuations, *Phys. Rev. Lett.* **106**, 20, p. 204501.

Donev, A., Garcia, A. L., de la Fuente, A., and Bell, J. B. (2011b). Enhancement of Diffusive Transport by Nonequilibrium Thermal Fluctuations, *J. of Statistical Mechanics: Theory and Experiment* **2011**, p. P06014.

Donev, A., Nonaka, A. J., Sun, Y., Fai, T. G., Garcia, A. L., and Bell, J. B. (2014b). Low Mach Number Fluctuating Hydrodynamics of Diffusively Mixing Fluids, *Communications in Applied Mathematics and Computational Science* **9**, 1, pp. 47–105.

Dorfman, J. R., Kirkpatrick, T. R., and Sengers, J. V. (1994). Generic long-range correlations in molecular fluids, *Annual Review of Physical Chemistry* **45**, 1, pp. 213–239.

Eyink, G. L. and Xin, J. (2000). Self-similar decay in the kraichnan model of a passive scalar, *J. Stat. Phys.* **100**, 3-4, pp. 679–741.

Hynes, J. T., Kapral, R., and Weinberg, M. (1979). Molecular theory of translational diffusion: Microscopic generalization of the normal velocity boundary condition, *J. Chem. Phys.* **70**, 3, p. 1456.

Majda, A., Timofeyev, I., and Vanden-Eijnden, E. (2006). Stochastic models for selected slow variables in large deterministic systems, *Nonlinearity* **19**, 4, p. 769.

Majda, A. J. and Kramer, P. R. (1999). Simplified models for turbulent diffusion: theory, numerical modelling, and physical phenomena, *Physics Reports* **314**, 4-5, pp. 237–574.

Mazur, P. and Bedeaux, D. (1974). Renormalization of the diffusion coefficient in a fluctuating fluid II, *Physica* **75**, pp. 79–99.

Papanicolaou, G. C. (1976). Some probabilistic problems and methods in singular perturbations, *Rocky Mountain J. Math* **6**, 4, pp. 653–674.

Schulz, B., Täuber, D., Friedriszik, F., Graaf, H., Schuster, J., and Von Borczyskowski, C. (2010). Optical detection of heterogeneous single molecule diffusion in thin liquid crystal films, *Physical Chemistry Chemical Physics* **12**, 37, pp. 11555–11564.

Usabiaga, F. B., Bell, J. B., Delgado-Buscalioni, R., Donev, A., Fai, T. G., Griffith, B. E., and Peskin, C. S. (2012). Staggered Schemes for Fluctuating Hydrodynamics, *SIAM J. Multiscale Modeling and Simulation* **10**, 4, pp. 1369–1408.

Vailati, A., Cerbino, R., Mazzoni, S., Takacs, C. J., Cannell, D. S., and Giglio, M. (2011). Fractal fronts of diffusion in microgravity, *Nature Communications* **2**, p. 290.

Vailati, A. and Giglio, M. (1997). Giant fluctuations in a free diffusion process, *Nature* **390**, 6657, pp. 262–265.

Zarate, J. M. O. D. and Sengers, J. V. (2006). *Hydrodynamic fluctuations in fluids and fluid mixtures* (Elsevier Science Ltd).

Chapter 6

Hard Sphere Simulation by Event-Driven Molecular Dynamics: Breakthrough, Numerical Difficulty, and Overcoming the issues

Masaharu Isobe

Graduate School of Engineering, Nagoya Institute of Technology, Nagoya, 466-8555, Japan

Abstract

Hard sphere/disk systems are among the simplest models and have been used to address numerous fundamental problems in the field of statistical physics. The pioneering numerical works on the solid–fluid phase transition based on Monte Carlo (MC) and molecular dynamics (MD) methods published in 1957 represent historical milestones, which have had a significant influence on the development of computer algorithms and novel tools to obtain physical insights. This chapter addresses the works of Alder's breakthrough regarding hard sphere/disk simulation: (i) event-driven molecular dynamics, (ii) long-time tail, (iii) molasses tail, and (iv) two-dimensional melting/crystallization. From a numerical viewpoint, there are serious issues that must be overcome for further breakthrough. Here, we present a brief review of recent progress in this area.

6.1 Introduction

The first clear evidence of the solid–liquid phase transition in the hard-sphere system was obtained in 1957 using a novel methodology "Molecular Dynamics (MD)" developed by Alder and Wainwright [Alder and Wainwright (1957)]. This groundbreaking paper and their subsequent work have provided new perspectives in many fields, eventually expanding into the new research fields of computational statistical physics and molecular simulation with the development of the computer. In 2007, Hiwatari and the author organized 50th anniversary celebration events in Kanazawa, Japan: "The 50th Anniversary of the Alder Transition" [Hiwatari and Isobe (2009)]; the event included two conferences. The first of these was a special session of 21st Annual Meeting of the Molecular Simulation Society of Japan, November 28, 2007, and the second was a symposium on "The 50th Anniversary of the Alder Transition—Recent Progress on Computational Statistical Physics—," November 29–30, 2007, in which 70 researchers participated. These events had a significant influence on the participants, especially young scientists, providing: (i) a historical viewpoint, (ii) an excellent opportunity to reconsider the possibility and promising future of molecular simulation, and (iii) a valuable opportunity to talk in front of the founder of molecular simulation to deepen their insight. The events were very successful. The published proceedings from these events include not only papers from the speakers but also several invited papers to celebrate the special occasion [Hiwatari and Isobe (2009)].

Professor Alder's work has had a tremendous influence on the author's research, especially after the author developed efficient algorithms for Event-Driven MD (EDMD) in hard disk systems as a graduate student at Kyushu University. On the way to the airport after the symposium in 2007, the author and Professor Alder discussed his presentation and agreed to collaborate on the molasses tail problem. Discussions continued via emails and visits to the Bay area by the author, ultimately resulting in the publication of three papers (see, Sec. 6.4). In 2008, Professor Alder arranged for Prof. Hiwatari and the author to visit a number of interesting places around the Bay area, including Lawrence Livermore National Laboratory (LLNL), Google, Stanford University, University of California, Berkeley, and the computer history museum. The same year, a United States–Japan bilateral workshop —Large-scale Molecular Dynamics Simulation and Related Topics— was organized at U.C. Berkeley. During the author's long stay at U.C. Berkeley in 2012, discussions were held nearly every week. Our families often shared meals; we enjoyed talking with Berni and his wife about various issues (child care, food, life in the USA, and literature). Given his impressive dedication to his work and wonderful outlook on life, Berni Alder is not only a collaborator but also a mentor and friend. The author would like to dedicate this chapter to Professor Berni Alder on the celebration of his 90th birthday.

This chapter addresses Alder's breakthrough works in the classical mechanics of the hard sphere/disk system from a numerical viewpoint. The chapter is organized as follows: simulation methodologies on the hard sphere/disk system are described in Sec. 6.2. The two-dimensional long-time tail and molasses tail problems are reviewed in Sec. 6.3 and Sec. 6.4, respectively. Sec. 6.5 focuses on the two-dimensional (2D) melting problem, where dislocations are calculated to estimate core energy by changing system size and methods. Concluding remarks are presented in Sec. 6.6.

6.2 Hard sphere/disk simulation

A many-body system with an impenetrable repulsive core (i.e., a hard disk in 2D systems or hard spheres in three-dimensional (3D) systems) and the associated phase transitions have long been investigated as one of the fundamental subjects in statistical physics. From a historical viewpoint, this simple microscopic model played a crucial role in understanding various macroscopic physical phenomena and provided an essential concept for constructing the relevant theory. Boltzmann constructed kinetic theory in the 19th century. Computer simulation methods for hard sphere/disk systems using Monte Carlo (MC) [Metropolis et al. (1953); Wood and Jacobson (1957); Krauth (2006)] and molecular dynamics (MD) [Alder and Wainwright (1957, 1959); Erpenbeck and Wood (1977); Rapaport (2004)] methodologies were implemented in the 20th century [Wood (1986)]. Following these pioneering works, along with the revolutionary development of computers, many algorithms have been developed [Allen and Tildesley (1987); Frenkel and Smit (2001)], which have since become powerful methodologies to clarify and visualize dynamic and cooperative aspects of dense liquid states and phase transitions at the atomic level [Hansen and McDonald (2006)].

In the hard sphere/disk model, the pair potential $\phi(r)$ for distance r from the center of the core is described as

$$\phi(r) = \infty \quad (r < 2\sigma), \tag{6.1}$$

$$= 0 \quad (r \geq 2\sigma), \tag{6.2}$$

where σ is the radius of the sphere/disk. The numerical algorithm of the hard sphere/disk model to solve its dynamics numerically is referred to as "Event-Driven MD (EDMD)" [Alder and Wainwright (1957, 1959); Erpenbeck and Wood (1977); Rapaport (2004)]. In the EDMD, the trajectories of the sphere/disk governed by Newton's classic equations of motion are successively generated. Next, pairs of collisions and collision times can be sorted in the list of potential collisions, which are estimated algebraically based on the updated positions and velocities. The evolution of hard disks/spheres evolved via events such as particle collision, wall collision, and sub-cell crossing as opposed to integrating differential equations of motion with a constant time step [Rapaport (2004)]. Although it appears straightforward, naive programming of this procedure results in complexity of $\mathcal{O}(N^2)$ per event in N particle systems. For better efficiency in large-scale systems, a special knowledge of data structure and scheduling arrangement of the queue of future events is required.

Rapaport (1980) [Rapaport (1980, 2004)] made a breakthrough with complexity becoming $\mathcal{O}(N \log N)$ using a binary search tree. In the 1990s, further progress was made in the associated algorithms by simplifying the data structure [Lubachevsky (1991); Marín et al. (1993, 1995)]. Marín and Cordero (1995) [Marín et al. (1995)] suggested that the complete binary tree constructed by local minima event time of each particle was the most efficient search method to detect the next event for all density regions. The author [Isobe (1999)] proposed further simplification for efficiency by utilizing exclusive particle grids and a dynamic upper cut-off time. In the 21st century, there has been a great deal of active research regarding the extension of EDMD to include non-spherical particles, as well as attempts to reduce calculation complexity [Donev et al. (2005); Paul (2007); Bannerman et al. (2011)].

Recently, a rejection-free hard sphere/disk ECMC algorithm was developed [Bernard and Krauth (2009)]. The simplest version of the ECMC algorithm is as given below. (i) A tagged sphere/disk α and direction of the trial (move) are determined by a random number. (ii) The sphere/disk α moves until it collides with another sphere/disk β along a straight line. (iii) The sphere/disk β moves in the same direction as α until it collides with another sphere/disk γ. (iv) This chain of collisions continues until total displacement of the sphere/disk regarding collisions equals a certain fixed length L_c. (v) Another tagged sphere/disk α' and direction of the trial (move) are sampled by a random number. (vi) The same collision process is repeated successively. If the moves are restricted only in the $+x$, $+y$, and $+z$ directions for efficiency, then the detailed balance is broken. However, it was shown to satisfy global balance and ergodicity [Michel et al. (2014)]. This algorithm was first applied to the 2D melting problem [Bernard and Krauth (2011); Engel et al. (2013)], but has recently been applied to the 3D melting problem [Isobe and Krauth (2015)]. In these studies, ECMC was shown to speed up equilibration for hard sphere/disk systems by several orders of magnitude compared with both EDMD [Alder and Wainwright (1959); Isobe (1999)] and local (conventional) Monte Carlo (LMC) [Metropolis et al. (1953)] methods. The concept of ECMC can be extended to continuous potentials and spin systems, in which the speed-up was also demonstrated [Bernard and Krauth (2012); Kapfer and Krauth (2013); Michel et al. (2014); Kapfer and Krauth (2014); Michael et al. (2015); Nishikawa et al. (2015)].

In hard sphere/disk systems above the melting transition point, a long relaxation time is required to obtain *equilibrium* for both positions and velocities of particles in the system as a whole. However, positional equilibration becomes difficult in dense systems as the excluded volume effect becomes dominant, and the particles become trapped in metastable states for a long time. The thermodynamic properties after equilibration generated by ECMC and EDMD are precisely the same [Engel et al. (2013); Isobe and Krauth (2015)]. As ECMC is much faster for equilibration of positions than EDMD in dense systems, ECMC is useful for equilibration of positions to reduce the CPU time in such dense systems. A more advanced method to improve the efficiency *Hybrid Scheme* was proposed [Isobe (2016)]. After generation of configurations by ECMC, the equilibrium could be easily obtained after velocity equilibration performed by short EDMD. Note that the *Hybrid scheme* between LMC and EDMD can be realized but is not suitable from the viewpoint of CPU time, as LMC is much slower than EDMD for positional equilibration [Engel et al. (2013); Isobe and Krauth (2015)]. For example, in time-dependent phenomena such as freely cooling granular gas, the production runs by EDMD can be initiated from the equilibrium configurations generated by the "*Hybrid Scheme*". This scheme can be applied to a relatively wide variety of physical phenomena regarding hard sphere/disk systems, including not only static (thermodynamic) but also time-dependent (non-equilibrium) dynamic properties. Possible future applications include jamming and glassy systems, as well as disordered systems.

6.3 Long-time tail problem

The transport coefficients derived from the integral form of the autocorrelation function of the time-dependent properties in linear response theory [Kubo et al. (1991); Nakano (1993)] (the so-called Green–Kubo expression) were first investigated numerically about 50 years ago by Alder and co-workers [Alder and Wainwright (1967, 1969, 1970); Alder et al. (1970)].

They discovered the power decay $\sim t^{-d/2}$ (t: time, d: dimension) of the velocity autocorrelation function (VACF) in moderately dense hard-sphere fluids with about 1000 hard disks, as opposed to ordinary exponential decay [Alder and Wainwright (1970)]. In the 2D case ($d = 2$), the diffusion coefficient shows logarithmic divergence or an absence of conventional hydrodynamics. This problem, known as the "2D long-time tail problem" has become the central issue of non-equilibrium statistical physics.

The diffusion coefficient D in the Green-Kubo expression [Kubo et al. (1991); Nakano (1993)] is described as

$$D \propto \int_0^\infty \langle v_x(t) v_x(0) \rangle dt, \tag{6.3}$$

where $v_x(t)$ is a tagged particle velocity at time t and $\langle \cdots \rangle$ indicates ensemble averages. Given that a tagged particle in the liquid state collides randomly with another particle, memory of the particle's initial velocity will be lost after several collisions under the assumption of molecular chaos, in which the time correlation function decays exponentially. However, numerical simulations show the following form of decay in liquids in a moderately dense system [Alder and Wainwright (1970)]:

$$\Phi(t) = \langle v_x(t) v_x(0) \rangle \sim \left(\frac{t}{t_0} \right)^{-\frac{d}{2}}, \tag{6.4}$$

where time t is divided by the mean free time t_0 for each density. In the 2D case ($d = 2$), the contribution from the algebraic form of the long-time tail in eq. (6.4) leads to logarithmic divergence of diffusion $D \sim \ln(\infty) + const.$ in eq. (6.3), in which hydrodynamics breaks down.

Numerical Difficulty: From a numerical viewpoint, several issues make it difficult to clarify the situation. (i) Sound wave propagation disturbs the pure VACF over a long time via periodic boundary conditions (PBCs); thus, a large system containing many particles is required. (ii) An accurate long-time correlation tail requires long-time simulation. (iii) To clarify the functional form of the VACF tail, it is necessary to average data with an enormous number of independent physical properties obtained by the simulation.

We revisited this problem [Isobe (2008, 2009)] by performing a large-scale, long-time, systematic EDMD simulation with one-million hard disks using a fast modern algorithm [Isobe (1999)]. To clarify the effects of PBCs,

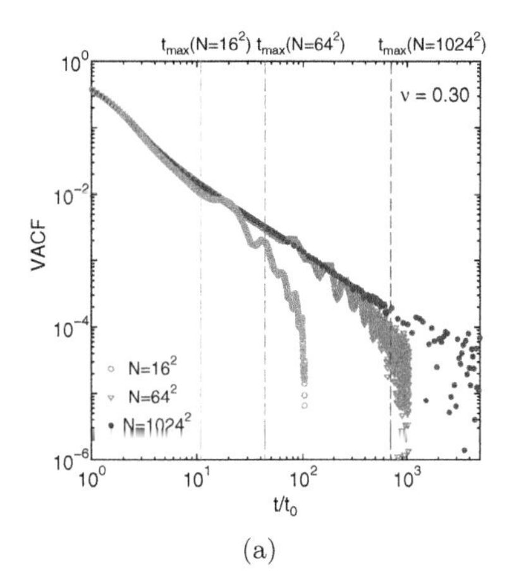

(a)

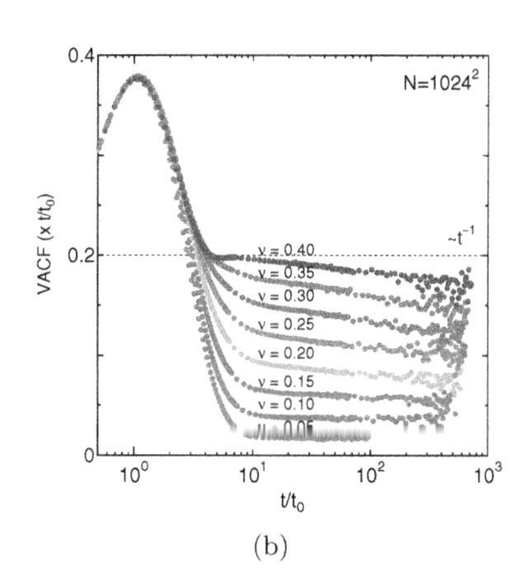

(b)

Fig. 6.1 (a) System size dependence of the VACFs at the packing fraction $\nu = 0.30$. (b) Packing fraction dependence of VACFs in a one-million hard disk fluid is shown. The vertical axis is multiplied by t/t_0 in the semi-log plot. (Figs. 2 and 3 in Ref. [Isobe (2008)])

the system-size dependence of the VACF of various sizes at a fixed packing fraction ($\nu = 0.30$) is shown in Figure 6.1 (a) (Fig. 2 in Ref. [Isobe (2008)]). Clear artificial disturbance due to sound wave propagation derived from PBCs in VACFs was found. For one million particles ($N = 1024^2$), we can discuss the pure VACF of the thermodynamic limit at more than $t/t_0 \sim 500$.

To investigate the functional form of the long-time tail, semi-log plots of the VACFs with different packing fractions are shown in Fig. 6.1 (b) (Fig. 3 in Ref. [Isobe (2008)]). The vertical axes are multiplied by t/t_0 to show the deviation from the conventional prediction of decay $(t/t_0)^{-1}$ more clearly. In the case of a dilute gas ($\nu = 0.05$), the tail of the VACF ($t/t_0 > 10$) approaches a flat line form (power decay α with $\alpha \sim -1$). In contrast, in a moderately dense gas ($\nu = 0.15 \sim 0.35$), the VACFs decay faster than the power form. Under the assumption of the power form of the long-time tail, we fitted the numerical data with a function of the form $\sim (t/t_0)^\alpha$ by changing the parameter α, i.e., $\alpha \sim -1.08, -1.06$, and -1.03 for $\nu = 0.18, 0.30$, and 0.45 respectively. Our numerical results for long-time tails in a moderately dense liquid were slightly faster than the inverse power form ($\sim 1/t$), which appeared to converge to the prediction curve based on self-consistent mode-coupling theory (MCT) ($\sim 1/(t\sqrt{\ln t})$) [Wainwright et al. (1971); Kawasaki (1971); Frenkel (1989)]. These phenomena have not been discussed in previous work [Alder and Wainwright (1970)] in which the true VACF could only be seen within the timescale $t/t_0 \sim 30$.

The physical origin of the long-time tail in VACF is recognized as the collective flows involving many particles surrounding a tagged particle. In the hydrodynamic model [Alder and Wainwright (1970)], a small element of fluid with an initial velocity in the momentum space of the equilibrium creates a compressible pressure to the front. During the evolution, the pressure relaxation causes a double vortex flow pattern, similar to the case of a particle moving in a compressible continuum fluid at the macroscopic level. The microscopic vortex flows push the tagged particle again, leading to the positive long-time tail of VACF. Alder and Wainwright [Alder and Wainwright (1969, 1970)] visualized these flow fields surrounding a tagged particle by EDMD. They investigated a system including 224 hard disks at $\nu \sim 0.4534 \cdots$ and a time difference $\tau/t_0 = 9.9$, in which 17 velocity vectors by EDMD were shown with macroscopic hydrodynamics. It was surprising that the hydrodynamics would apply quantitatively at the atomic level.

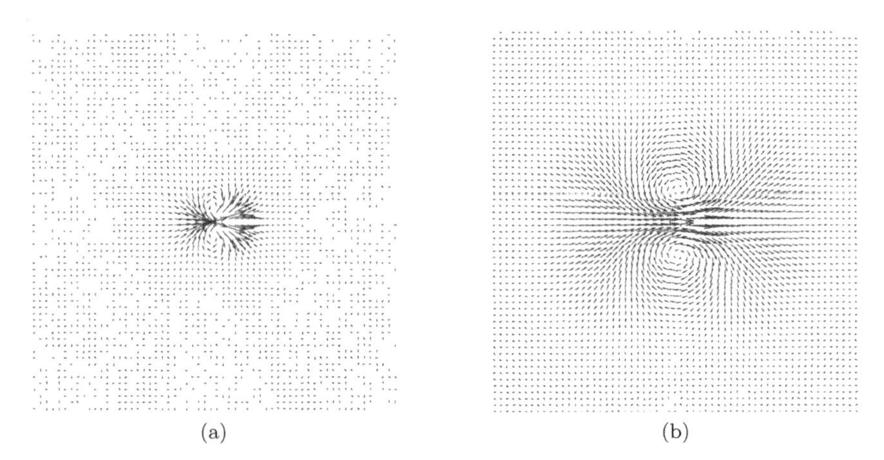

Fig. 6.2 Flow patterns for (a) $\tau/t_0 = 3$ and (b) $\tau/t_0 = 10$ around the right-moving tagged particle, located at the center of the figures. The packing fraction was set at $\nu = 0.45$. (Figs. 3 and 4 of Ref. [Isobe (2009)])

To visualize such a flow pattern, the velocity field was also revised by the author [Isobe (2009)]. Figure 6.2 shows the velocity field around a tagged particle (located at the center of the figure) at $(N, \nu) = (32^2, 0.45)$ with high resolution (grid size: $\sim 0.70\sigma$) for sufficiently long runs (Figs. 3 and 4 of Ref. [Isobe (2009)].); the time differences are fixed at (a) $\tau/t_0 = 3$ and (b) 10, respectively. If the particle moves randomly under the assumption of molecular chaos at the microscopic level, then there is no current or correlation around the tagged particles. Two clear vortexes for both sides of the evolving tagged particle were created, which grew gradually over time. Our calculation also provided clear evidence of hydrodynamic flow as the origin of the long positive tail.

6.4 Molasses tail problem

The slow dynamics of supercooled liquids and the glass transition have been investigated intensively for several decades [Berthier et al. (2011); Chandler and Garrahan (2010)]. Density fluctuations in the glass have been studied actively; however, the microscopic mechanism of the stress field relaxation has not been clarified. The shear stress autocorrelation function (SACF) is defined by $\langle J_{xy}(t)J_{xy}(0)\rangle$, where $J_{xy}(t)$ is the momentum flux at time t. The Green–Kubo expression for the viscosity is as follows:

$$\eta \propto \int_0^\infty \langle J_{xy}(t)J_{xy}(0)\rangle dt. \tag{6.5}$$

The decay of SACF has been investigated by MCT and kinetic theory, both leading to the same power form as in the case of the VACF, as shown in Sec. 6.3. However, numerical studies in the 1980s showed that the tail of the amplitude in the potential part of SACF of dense fluids was several orders of magnitude greater than that predicted by theory. This long time correlation in the potential part of the SACF has been called the "molasses tail" to differentiate it from the hydrodynamic origin of the long-time tail in the VACF and to emphasize its relation to the highly viscous glassy state [Alder (1986)]. The discrepancy was first suggested to be due to insufficient numerical simulation to resolve the long tail. However, this tail was eventually considered to be due to slow structural relaxation in the dense liquid around the first peak of the structure factor rather than the hydrodynamic origin at a longer length, with which the mode coupling theory of glass transition was proposed.

The momentum current J_{xy} can be decomposed into

$$J_{xy} = J_{xy}^K + J_{xy}^P, \tag{6.6}$$

where J_{xy}^K and J_{xy}^P are the kinetic and potential parts of the momentum current, respectively. Three types of correlation occur, i.e., kinetic-kinetic, potential-potential, and cross kinetic-potential parts. The potential part of the SACF is given by

$$\rho(t) = \langle J^P_{xy}(t) J^P_{xy}(0) \rangle, \tag{6.7}$$

in which the molasses tail appears.

Numerical Difficulty: In the late 1980s, Ladd and Alder speculated that the long-time tail of SACF in dense liquid was caused by transient crystal nuclei [Ladd and Alder (1989)]. They performed EDMD simulations and focused on the observation that the potential part of the SACF and the angular orientational autocorrelation function (OACF) are identical in the long time limit and show non-algebraic decay in time. To confirm that the microscopic origin of the non-algebraic decay comes from structural relaxation rather than hydrodynamic flow, they attempted to understand it by decomposing the OACFs into two-, three-, and four-body correlations; however, these correlations were not obtained accurately due to computer limitations. Calculation of the four-body time correlation functions is computationally time-consuming.

Estimation of the cooling rate to prevent crystallization requires information on the growth rate of crystal nuclei and the associated transient time, which can only be obtained from the decomposed time correlation functions. Analyses of the slow-decaying OACF $C(t) = \langle O_{xy}(t) O_{xy}(0) \rangle$ and its decomposition are expected to be key factors in understanding the onset of supercooled liquids. The author and Professor Alder [Isobe and Alder (2009, 2010, 2012)] revisited this problem. First, we followed the slow decay of the pair part of OACF in two dimensions, consisting of hard disks near the solid–fluid transition point as outlined previously Ref. [Ladd and Alder (1989)], placed in a square box with periodic boundary conditions, using efficient EDMD [Isobe (1999)].

$$O_{xy}(t) = \sum_c \frac{x_{ij} y_{ij}}{(2\sigma)^2} \delta(t - t_c), \tag{6.8}$$

where $\sum_c$ is the contribution at collision time t_c at which $(x_{ij}, y_{ij}) = (x_i - x_j, y_i - y_j)$ are the relative positions of the sphere/disk between i and j. To avoid the delta function singularity of $O_{xy}(t)$, the alternative Einstein–Helfand expression involving the second derivative is adopted to calculate the correlation function.

The total correlation function $C(t)$ can be decomposed into pair $C_2(t)(ij - ij)$, triplet $C_3(t)$ $(ij - ik)$, and quadruplet $C_4(t)$ $(ij - kl)$ contributions [Ladd et al. (1979)], where i, j, k, l are particle indices.

$$C(t) = C_2(t) + C_3(t) + C_4(t). \tag{6.9}$$

For example, the pair contribution $C_2(t)$ is defined as

$$C_2(t) = \frac{1}{N}\left\langle \sum_{i}^{N} \sum_{j(j>i)}^{N} O_{xy}^{ij}(t) O_{xy}^{ij}(0) \right\rangle. \tag{6.10}$$

Near the solid–fluid phase transition, three regimes in the relaxation of C_2 were identified, i.e., the kinetic, molasses (stretched exponential), and diffusional power decay [Isobe and Alder (2009)]. The origin of the stretched exponential decay at intermediate times due to the presence of crystal clusters of various sizes decaying at different rates was confirmed by detection using the bond-orientational order parameter [Halperin and Nelson (1978)]. We then focused on the rapidly growing time of the OACF decay in dense systems. This established the length of time for which the largest nuclei existed at each density. The largest cluster near the freezing density was only a few sphere diameters in size and persisted for only about 30 ps for the typical argon parameters.

The theoretical predictions for the final power law decay were also compared with numerical results [Isobe and Alder (2010)].

To make further progress quantitatively, the OACF of the quadruplet component, $C_4(\Delta R, t)$, as a function of the distance between the two colliding pairs ΔR, was considered. Based on C_4, it is possible to quantify how the size distribution of clusters changes regarding both time and density, which can be evaluated to determine how fast the density must increase to obtain kinetically glassy states instead of crystallization. Resolution of C_4 is computationally time-consuming; thus, we developed two methodologies [Isobe and Alder (2012)]. The first is a more efficient algorithm for calculating C_2 and C_4 contributions to the OACFs by coarse-grain (CG) detection of neighbors rather than the collision-based calculation. Figure 6.3(a) shows a comparison of C_2 at $\nu = 0.65$ and 0.69 between collision-based and CG methods, the results of which showed fairly good agreement with each other. The other was an extension of the bond-orientational order parameter ϕ_6^i to further neighboring shells [Halperin and Nelson (1978)]. The bond-orientational order parameter is organized based on the angle between the position vector from disk J to I and "an arbitrary fixed reference axis." To consider the angle between first and second neighbor pairs (and more), the bond angle can be redefined by relative vectors between

I-J and I-K (etc.). The complex generalized order parameter (GOP) for a tagged particle I with the actual position vectors $\mathbf{r}_{JI}$ can be generalized by

$$\Phi_s^I = \frac{1}{N_I(N_I - 1)} \sum_{J \neq J'}^{N_I} \chi_s(\mathbf{r}_{JI}, \mathbf{r}_{J'I}), \tag{6.11}$$

$$\chi_s(\mathbf{r}_{JI}, \mathbf{r}_{J'I}) = \exp\left(\mathrm{i}s\theta(\mathbf{r}_{JI}, \mathbf{r}_{J'I})\right), \tag{6.12}$$

$$\theta(\mathbf{r}_{JI}, \mathbf{r}_{J'I}) = \theta_I^{JJ'} = \cos^{-1}\left(\frac{\mathbf{r}_{JI} \cdot \mathbf{r}_{J'I}}{|\mathbf{r}_{JI}||\mathbf{r}_{J'I}|}\right). \tag{6.13}$$

If $s = 6$ and $\mathbf{r}_{J'I} = (1, 0)$ (i.e., unit vector of x-axis) are fixed, then we can reduce the above expression to the usual ϕ_6 order parameter described.

In Fig. 6.3(b), the spatial distribution of the GOP for third neighbors Φ_{18}^i in a 4096 disk system at $\nu = 0.69$ is shown.

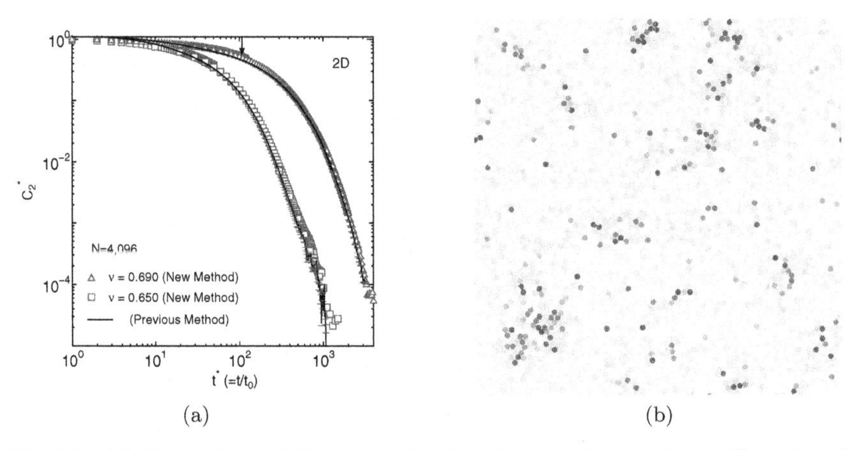

Fig. 6.3 (a) Comparison of C_2 at two densities between the previous collision-based method and the new method. The vertical axis is normalized and the horizontal axis is the time scaled by the mean free time, t_0. (Fig. 6 in Ref. [Isobe and Alder (2012)]) (b) The spatial distribution of the generalized order parameter for third neighbors Φ_{18}^i in a 4096 particle system at a given time at $\nu = 0.69$. (Fig. 9 (b) in Ref. [Isobe and Alder (2012)]) The darker the region, the closer Φ_s^i is to unity.

With these higher order parameters, we successfully demonstrated that the novel results were in good agreement with those obtained with the previous method. However, the algorithm was two orders of magnitude faster, allowing calculation of two different pairs of particles as a function of their separation $C_4(\Delta R, t)$, which can be used to characterize the size of the transient crystals.

Table 6.1 Relaxation time by various correlation functions and order parameters is summarized.

ν	$\tau(C_{total})$	$\tau(C_2)$	$\tau_1(C_4)$	$\tau_2(C_4)$	$\tau^1_{\rm GOP}$	$\tau^2_{\rm GOP}$	$\tau^3_{\rm GOP}$
0.69	12	133	78	66	40	20	15
0.65	6	41	21	21	9	7	6

The lifetimes of these clusters given in units of t_0 (mean free time) are determined from the autocorrelations for $C_{total}(t)$, $C_2(t)$, $C_4(\Delta R, t)$, and $C_{\rm GOP}(t)$ that are of the stretched exponential form. In Table 6.1, the relaxation time τ for C_{total}, C_2, and τ_k of $C(\Delta R, t)$ for k-th peaks, and $\tau^s_{\rm GOP}$ for n-th N.N. at $\nu = 0.69$ and 0.65 are summarized. As expected, the relaxation time increases when the packing fraction increases and decreases for higher neighbor shells. The lifetime and size of transient crystal nuclei in the pre-freezing phase were clarified by these methods. The slow decay of SACF near freezing is caused by a large number of small transient crystal nuclei, which lead to the molasses tail. The rapid increase in the relaxation time of clusters with density, as well with their number and size, is closely related to the rapid increase in viscosity near freezing. If we cool the system or compress much faster than the relaxation time estimated by the methods, then conditions under which glass may form can be estimated. The potential part of the SACF is the central issue of jamming; these tools will clarify the properties in densely packed systems.

6.5 Two-dimensional melting

6.5.1 *Long debate*

In 1962, Alder and Wainwright [Alder and Wainwright (1962)] showed numerically that 2D hard disk systems occur during the phase transition between the liquid state and crystal solid by increasing the packing fraction (density). For more than 50 years, there has been active debate regarding how the melting transition occurs in this system, i.e., whether a first order [Alder and Wainwright (1962); Alder et al. (1968); Chui (1982)] or a Kosterlitz–Thouless-type continuous order phase change occurs between liquid and crystal solid phases via a "hexatic phase", known as *KTHNY theory* [Kosterlitz and Thouless (1973); Halperin and Nelson (1978); Nelson and Halperin (1979); Young (1979); Nelson (2002)].

Numerical Difficulty: This debate is difficult to resolve numerically, because simulation requires not only larger particle systems than the correlation length of the order parameter but also longer equilibration time to obtain true equilibrium states. Although hard sphere/disk systems are simple, the relaxation time of positions and coarsening process around the transition point are slow due to the high activation free energy for crystallization. Computer resources place a limitation on large-scale simulations. Furthermore, the timescale is also longer in the coexistence regions; this is due to the excess free energy of surface tension regarding coexisting phases in a system of finite size. This lack of either system size or equilibration time caused disagreement between many studies. Novel specialized, efficient algorithms are required to overcome these problems.

Recently, extensive large-scale simulations in 2D one million hard disk systems in the region of the phase transition have been performed using three state-of-the-art algorithms [Engel et al. (2013)], i.e., ECMC [Bernard and Krauth (2009)], Massive Parallel Monte Carlo [Anderson et al. (2013)], and EDMD [Isobe (1999)]. These simulations independently reproduced the Mayer–Wood loop in the equation of state [Mayer and Wood (1965)] with high precision. The results were consistent with those obtained previously with ECMC [Bernard and Krauth (2011)], which showed novel melting scenarios for the first-order transition between hexatic and liquid phases via a coexistence phase and continuous transition between hexatic and crystal solid phases. These melting scenarios have been confirmed in soft disk systems [Kapfer and Krauth (2014)].

6.5.2 *Defects and core energy*

From the viewpoint of the microscopic melting process, the static distributions and dynamic behavior of isolated/cluster dislocations should be helpful to understand the whole picture of this scenario. Here, we focus on the distribution of "topological defects" in which the number of fractions of topological defects as a function of packing fraction ν are investigated. In 2D perfect crystal solids, a disk has 6 nearest neighbors (N.N.). We define a "defect" as a disk *not* having 6 N.N. "Disclination" and "dislocation" are defined as isolated defects with 5 or 7 N.N. and isolated bound defects with 5 and 7 N.N. pairs, respectively. KTHNY theory suggests a two-step melting scenario in which the unbinding of dislocation pairs ($5-7-5-7$ quartets) into free dislocations causes the solid-hexatic transition, and the dissociation of dislocations ($5-7$ pairs) into free disclinations induces the hexatic–liquid transition. After publication of the KTHNY theory, Chui proposed another theory in which the 2D melting scenario occurred via the spontaneous appearance of string-like grain boundaries when the dislocation core energy E_c is less than $2.84k_BT$ [Chui (1982)]. Saito followed the proposal of Chui, incorporating MC lattice model simulations [Saito (1982)] and reported string-forming defects for $E_c < 2.84k_BT$ as grain boundaries revealing a first-order melting transition caused by dislocations, while for $E_c > 2.84k_BT$ a KTHNY melting transition occurs.

In the actual simulation in continuous space, it is difficult to identify isolated defects as they are distributed as more complicated configurations rather than isolated simple unbound defect pairs. The reasons can be summarized as follows [Qi et al. (2010)]. (i) In the highly concentrated regions of defects, it is difficult to determine whether a defect has disappeared or not by the presence of neighboring defects. (ii) Defects are usually organized as large clusters, which may also be part of a cluster of dislocations or disclinations. (iii) Particles with 4 or 8 N.N. cannot be categorized as dislocations or disclinations. The above factors make it difficult to accurately determine the number of isolated dislocations or disclinations.

To avoid overestimation of the fraction of isolated dislocations, it is necessary to exclude transient dislocations due to thermal fluctuation. It has been noted that excluding such effects is important with regard to accurate estimation of the core energy [Qi et al. (2010)]. The inherent structure is often considered to be realized at zero temperature. Transient dislocations disappear when we consider the inherent structure and only the intrinsic structure is investigated. In the crystal solid phase, there are no dislocations or grain boundaries in the inherent structure.

In previous studies, the core energy E_c was measured roughly from the Boltzmann distribution of the concentration of isolated dislocations [Tobochnik and Chester (1982); Nosenko et al. (2009); Qi et al. (2010, 2014)].

$$\frac{n_d}{1 - n_d} = \exp\left(-\frac{E_c}{k_BT}\right), \tag{6.14}$$

where n_d is the concentration of isolated dislocations. With this formula based on the concentration of isolated defects, we can estimate the core energy with the original configuration (i.e., non-coarse-grained) obtained by the simulation. In this case, *non* 6 N.N. includes thermal fluctuations, dislocation pairs, and defect clusters. As mentioned above, this led to unexpected effects, i.e., overestimation of dislocations, resulting in underestimation of the core energy. In following subsection, we consider coarse-grained disk position by δt to eliminate unstable dislocation with thermal fluctuation.

To investigate the nature of transition, it is important to quantitatively estimate dislocations and core energy from the configurations obtained by simulations. For this purpose, it is necessary to clarify two effects: (i) the effects of thermal fluctuation and (ii) the effects of choosing methods for N.N. detection. Both cause unusual over/underestimation of dislocations, which directly affects the core energy. Here, we examine the two effects with a focus on system size dependence.

6.5.3 *Thermal fluctuation*

In a dense many-body system with repulsive force, a particle is surrounded by neighboring particles and fluctuates within the cage formed by these neighbors. Under these conditions, the particle motion is composed of two parts, configurational and vibrational contributions (i.e., inherent structure and thermal fluctuation). To obtain a stable configuration, the inherent structure is often used as the potential energy minimum of the configuration space [Stillinger (2015)]. Other methods to identify the general potential to reduce the contribution of thermal fluctuation include steepest descent, conjugate gradient, and FIRE methods [Bitzek (2006)]. In the hard disk system, averaged positions $\overline{\mathbf{r}_i(t)}$ with a certain time δt are estimated as follows,

$$\overline{\mathbf{r}_i(t)} = \frac{1}{\delta t} \int_0^{\delta t} \mathbf{r}_i(t + t')dt'. \tag{6.15}$$

We only use $\delta t/t_0 = 100$ (t_0 is mean free time at $\nu = 0.720$) as we focus on dense systems in the region of $\nu = 0.698 - 0.722$. This time scale is sufficiently larger than that of thermal fluctuations (an order of a few mean free time t_0) and smaller than that of particle diffusion from the cage. It was confirmed that coarse-grained time scales $\delta t/t_0 = 100$ are smaller than typical diffusion time scales $\sim 10^3 \sim 10^4$ around the solid–fluid transition points.

In Fig. 6.4, typical trajectories of a tagged particle in a certain direction and its positions averaged over the time scale $\delta t/t_0 = 100, 500, 1000$ at $\nu = 0.708$ are shown. Trajectories with $\delta t/t_0 = 100$ actually suppress microscopic fluctuations, reducing noise and overestimation of defects via accidental fluctuation.

6.5.4 *Detection of neighbors*

Identification of N.N. in a many-particle system is a difficult task as there is no unique definition [Meel et al. (2012)]. Two methods are frequently used for this purpose, i.e., a fixed-distance cutoff and a Voronoi construction.

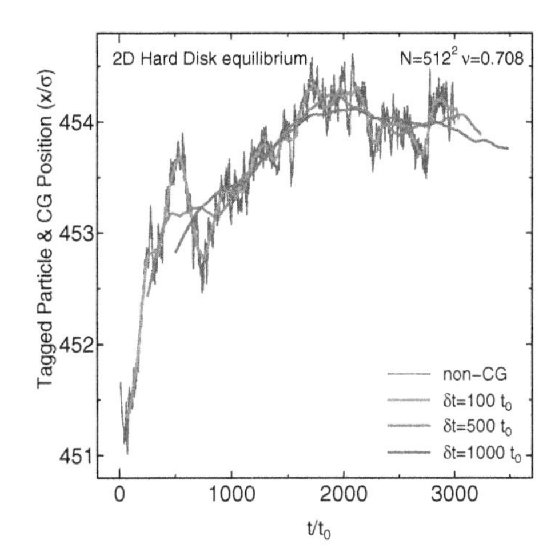

Fig. 6.4 Typical trajectories of a tagged particle in the x direction and its positions averaged over time scale $\delta t/t_0 = 100, 500, 1000$ in $(N, \nu) = (512^2, 0.708)$.

In the fixed distance cutoff method, one parameter, "cut-off distance," is needed, which is often used as the first minimum of the radius distribution function in a monodisperse homogeneous system. In the Voronoi construction, computational costs are relatively high; also, this method is known to be sensitive to thermal fluctuation, often leading to overestimation of N.N. Recently, Meel et al. developed a novel, efficient algorithm to overcome the problems of the above two methods [Meel et al. (2012)], called the Scale-free Algorithm for Nearest Neighbor (SANN). SANN introduces the concept of *solid angle* in three dimensions. This algorithm has significant advantages: (i) it can be applied to polydisperse and inhomogeneous systems; (ii) it is stable against thermal fluctuation; (iii) it is not necessary to choose parameters; and (iv) it has low computational costs.

Recently, we extended the scheme of SANN to 2D configurations and applied it to binary mixture hard disk systems (See Ref. [Isobe (2016)] for details). Although there is no explicit form of inequality of SANN in the 2D case, calculation using the bisection method to solve nonlinear equations worked fairly well without excessive computational costs. This method has the advantage of estimating N.N. even for polydisperse and/or inhomogeneous systems. Note that SANN cannot deal with next nearest neighbor correctly; thus, we considered a more general method to consider further neighbors (i.e., second and third neighbors) [Isobe and Alder (2012)] as described in Sec. 6.4.

6.5.5 *Spatial configurations of dislocation*

This subsection describes a comparison of spatial configurations of dislocation between the methods for detection of N.N. Figures 6.5 show the typical configurations of defects in $N = 256^2$ using time-averaged positions with $\delta t/t_0 = 100$ described in Sec. 6.5.3. All systems are equilibrated completely by a sufficiently long simulation of ECMC; thus, we focus on the packing fractions for liquid $\nu = 0.698$, hexatic $\nu = 0.718$, and solid $\nu = 0.722$. In Figs. 6.5(a)-(c), N.N. are detected by a fixed distance cut-off method, while in Figs. 6.5(d)-(f), N.N. are detected by the SANN method (see, Sec. 6.5.4). Each disk is drawn with the N.N. number detected by both methods, indicated by color as follows: 4 (orange), 5 (pink), 6 (green), 7 (blue), and 8 (dark blue). As discussed in Ref. [Meel et al. (2012)], SANN underestimates the average number of N.N. compared with fixed distance cut-off or Voronoi construction. As about half of the neighbors located around the first minimum of the radial distribution function are categorized as next nearest neighbors in the SANN method, there are many 5 N.N. (pink disks) in the configurations detected by SANN compared with the fixed distance cut-off method. Calculation of fraction distributions of defects for 4, 5, 7, and 8 neighbors confirmed that the number of neighbors 5 of SANN is much larger than that of the fixed distance cut-off method. In the fixed distance cut-off method, the numbers of 5 and 7 are of the same order for all densities, which seems to well describe the existence of the dislocation scenario. The numbers of 4 and 8 neighbors are negligible compared with 5 and 7 neighbors. The difference in ratio of N.N. number between methods is relatively large, which is one source of difficulty in accurately estimating the core energy.

6.5.6 *Core energy*

The core energy E_c is then estimated by changing the system size and methods. Based on eq. (6.14), the core energy is calculated in $N = 32^2$, 256^2, and 512^2 around the regions of $\nu = 0.698 \sim 0.722$ by two neighbor detection methods. As shown in Fig. 6.6, we found that all E_cs are increasing functions with regard to packing fraction. With CG configurations (i.e., averaged positions with $\delta t/t_0 = 100$), almost the same results are obtained between $N = 256^2$ and 512^2. In contrast, the core energy of $N = 32^2$ has larger values than those of $N = 256^2$ and 512^2. These results confirm that the system size $N = 32^2$ is not sufficiently large to investigate the 2D melting problem as suggested previously in Ref. [Bernard and

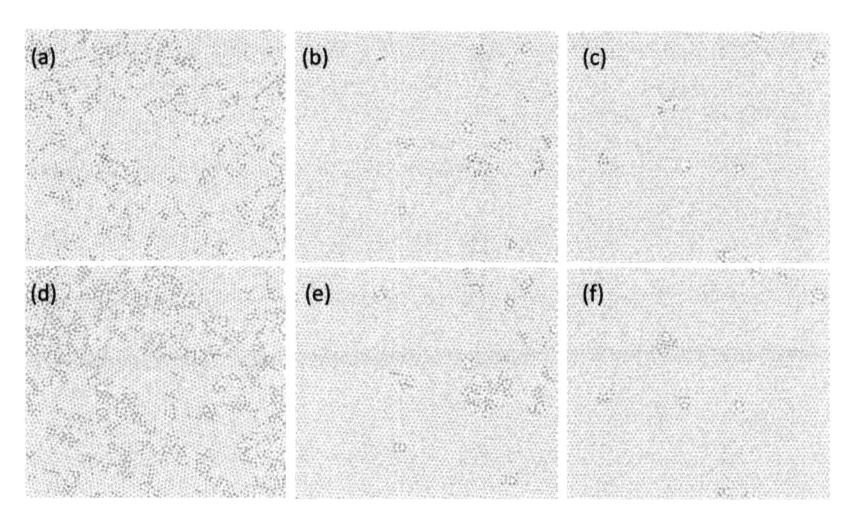

Fig. 6.5 Typical configurations of number of N.N. for (a)(d) $\nu = 0.698$, (b)(e) 0.718 and (c)(f) 0.722 in $N = 256^2$ are shown, respectively. Each disk is colored by the number of N.N., which are detected by a fixed distance cut-off method for (a)-(c) and the SANN method for (d)-(f).

Krauth (2012); Engel et al. (2013); Isobe (2016)]. We also found that the core energy is strongly dependent on the fluctuation of disks. With non-CG configuration including thermal fluctuations, the core energy has lower values than CG configurations, because thermal fluctuation causes overestimation of defects. As shown above, the core energy is sensitive to changes in both the neighbor detection method and thermal fluctuation. The most interesting point in Fig. 6.6 is that the values of core energy cross upward through the horizontal threshold line at $E_c = 2.84$. This value $E_c = 2.84$ is a critical point where the transition changes from a grain boundary scenario ($E_c < 2.84$) to a KTHNY scenario ($E_c > 2.84$). In SANN with CG configuration, the transition changes at packing fraction $\nu = 0.716$, consistent with a first-order melting scenario (i.e., grain boundary) between liquid-hexatic and continuous melting (KTHNY) between the hexatic-solid phase [Bernard and Krauth (2011); Engel et al. (2013); Qi et al. (2014)].

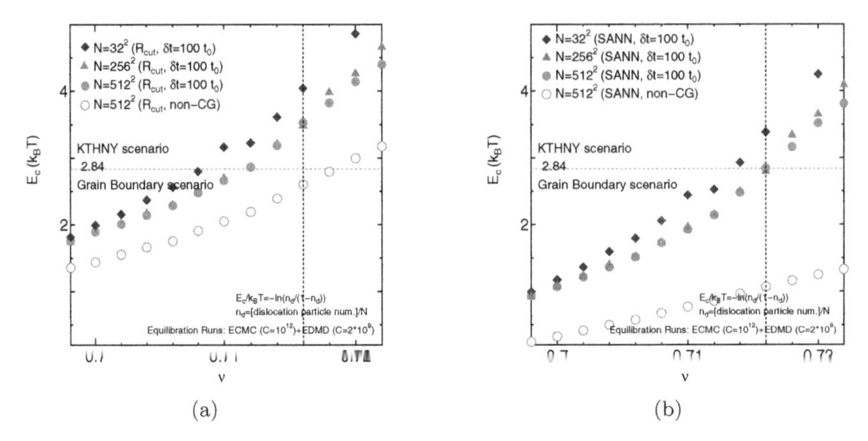

Fig. 6.6 The values of core energy with regard to packing fraction are shown for both (a) a fixed distance cut-off and (b) SANN methods

6.6 Concluding remarks

This chapter reviewed the works associated with Alder's breakthrough regarding the hard sphere/disk simulation, including EDMD, long-time tail, molasses tail, and 2D melting problems. These topics required overcoming serious numerical difficulties. Recent progress on these problems was also reviewed. With modern methodologies, the 3D melting/crystallization problem described in the Introduction has also been examined recently with much larger systems, up to one-million hard spheres [Isobe and Krauth (2015)]. The equation of state was reproduced where long-time simulations partially realized the coexistence configurations in the microcanonical ensemble finite-size box for larger systems. We found that ECMC was much faster than not only conventional MC [Metropolis et al. (1953)] but also EDMD, especially at high packing fractions.

The author would like to dedicate this chapter to Professor Berni Alder on the occasion of his 90th birthday.

Acknowledgements

The author is grateful to Profs. B. J. Alder, W. G. Hoover and W. Krauth for stimulating and fruitful discussions on this topics. He also grateful to Dr. K. Niki for helpful comments for improving the manuscript. This study was supported by JSPS Grant-in-Aid for Scientific Research No. 26400389. Part of the computations were performed using the facilities of the Supercomputer Center, ISSP, Univ. of Tokyo.

Bibliography

Alder, B. J. and Wainwright, T. E. (1957). Phase Transition for a Hard Sphere System, *J. Chem. Phys.* **27**, pp. 1208–1209.

Alder, B. J. and Wainwright, T. E. (1959). Studies in Molecular Dynamics I. General Methods, *J. Chem. Phys.* **31**, pp. 459–466.

Alder, B. J. and Wainwright, T. E.(1962). Phase Transition in Elastic Disks, *Phys. Rev.* **127**, pp. 359–361.

Alder, B. J., Hoover, W. G. and Young, D. A.(1968). Studies in Molecular Dynamics. V. High Density Equation of State and Entropy for Hard Disks and Spheres, *J. Chem. Phys.* **49**, pp. 3688–3696.

Alder, B. J. and Wainwright, T. E.(1967). Velocity Autocorrelations for Hard Spheres, *Phys. Rev. Lett.* **18**, pp. 988–990.

Alder, B. J. and Wainwright, T. E.(1969). Enhancement of Diffusion by Vortex-Like Motion of Classical Hard Particles, *J. Phys. Soc. Jpn. Suppl.* **26**, pp. 267–269.

Alder, B. J. and Wainwright, T. E.(1970). Decay of the Velocity Autocorrelation Function, *Phys. Rev. A* **1**, pp. 18–21.

Alder, B. J., Gass, D. M. and Wainwright, T. E.(1970). Studies in Molecular Dynamics. VIII. The Transport Coefficients for a Hard Sphere Fluid, *J. Chem. Phys.* **53**, pp. 3813–3826.

Alder, B. J.(1986). Molecular-Dynamics Simulations, in *Proceedings of the International School of Physics $\ll$ Erinco Fermi $\gg$ Course XCVII, Varenna on Lake Como, 1985*, G. Ciccotti and W. G. Hoover (eds.), (North-Holland, Amsterdam, 1986), pp. 66–80.

Allen, M. P. and Tildesley, D. J.(1987). *Computer Simulation of Liquids*, (Oxford University Press)

Anderson, J. A., Jankowski, E., Grubb, T. L., Engel, M. and Glotzer, S. C.(2013). Massively parallel Monte Carlo for many-particle simulations on GPUs, *J. Comput. Phys.* **254**, pp. 27–38.

Bannerman, M. N., Sargant, R., and Lue L.(2011). DynamO: A Free O(N) General Event-Driven Molecular Dynamics Simulator, *J. Comput. Chem.* **32**, pp. 3329–3338.

Bernard, E. P., Krauth, W., and Wilson, D. B.(2009). Event-chain Monte Carlo algorithms for hard-sphere systems, *Phys. Rev. E* **80**, 056704.

Bernard, E. P. and Krauth, W.(2011). Two-Step Melting in Two Dimensions: First-Order Liquid-Hexatic Transition, *Phys. Rev. Lett.* **107**, 155704.

Bernard, E. P. and Krauth, W.(2012). Addendum to "Event-chain Monte Carlo algorithms for hard-sphere systems", *Phys. Rev. E* **86**, 017701.

Berthier, L., Biroli, G., Bouchaud, J. -P., Cipelletti, L. and Saarloos W. Van. (eds.)(2011). *Dynamical Heterogeneities in Glasses, Colloids, and Granular Media*, (International Series of Monographs on Physics, Oxford Univ Press)

Bitzek,E., Koskinen, P., Gähler, F., Moseler, M. and Gumbsch, P.(2006). Structural Relaxation Made Simple, *Phys. Rev. Lett.* **97**, 170201.

Chandler, D. and Garrahan, J. P.(2010). Dynamics on the way to forming glass: bubbles in space-time, *Annu. Rev. Phys. Chem.* **61**, pp. 191–217.

Chui, S. T.(1982), Grain-Boundary Theory of Melting in Two Dimensions, *Phys. Rev. Lett.* **48**, pp. 933–935.

Donev, A., Torquato, S. and Stillinger, F. H.(2005). Neighbor list collision-driven molecular dynamics simulation for nonspherical hard particles. I. Algorithmic details, *J. Comput. Phys.* **202** pp. 737–764.

Engel, M., Anderson, J. A., Glotzer, S. C., Isobe, M., Bernard, E. P. and Krauth, W.(2013). Hard Disks Equation of State: First-Order Liquid-Hexatic Transition in Two Dimensions with Three Different Simulation Methods, *Phys. Rev. E* **87**, 042134.

Erpenbeck, J. J. and Wood, W. W. (1977). Molecular Dynamics Techniques for Hard-Core Systems, in *Modern Theoretical Chemistry Vol. 6, Statistical Mechanics Part B*, B.J. Berni (eds.), (Plenum, New York), **Chap. 1**, pp. 1–40.

Frenkel, D. and Ernst, M. H.(1989). Simulation of diffusion in a two-dimensional lattice-gas cellular automaton: a test of mode-coupling theory, *Phys. Rev. Lett.* **63**, pp. 2165–2168.

Frenkel, D.and Smit, B.(2001). *Understanding Molecular Simulation, From Algorithms to Applications*, (Academic Press, 2nd Edition)

Halperin, B. I. and Nelson, D. R.(1978). Theory of Two-Dimensional Melting, *Phys. Rev. Lett.* **41**, pp. 121–124.

Hansen, J. P. and McDonald, I. R.(2006) *Theory of Simple Liquids*, (Academic Press, London, 3rd Edition)

Hiwatari, Y. and Isobe, M. (eds.) (2009). The 50th Anniversary of The Alder Transition —Recent Progress on Computational Statistical Physics—, *Prog. Theor. Phys. Suppl. (Kyoto)* **178**.

Isobe, M.(1999). Simple and efficient algorithm for large scale molecular dynamics simulation in hard disk systems, *Int. J. Mod. Phys. C* **10**, pp. 1281–1293.

Isobe, M.(2008). The Long Time Tail of the Velocity Autocorrelation Function in a Two Dimensional Moderately Dense Hard Disk Fluid, *Phys. Rev. E* **77**, 021201.

Isobe, M.(2009). Vortex Flows in Two Dimensions — The Origin of Hydrodynamic Tail —, *Prog. Theor. Phys. Suppl.* **178**, pp. 72–78.

Isobe, M. and Alder, B. J. (2009). Molasses Tail in Two Dimensions, *Mol. Phys.* **107**, pp. 609–613.

Isobe, M. and Alder, B. J. (2010). Study of transient nuclei near freezing, *Prog. Theor. Phys. Suppl.* **184**, pp. 437–448.

Isobe, M. and Alder, B. J. (2012). Generalized Bond Order Parameter to Characterize Transient Crystals, *J. Chem. Phys.* **137**, 194501.

Isobe, M. and Krauth, W.(2015). Hard-Sphere Melting and Crystallization with Event-Chain Monte Carlo, *J. Chem. Phys.* **143**, 084509.

Isobe, M.(2016). Hard sphere simulation in statistical physics – methodologies and applications, *Mol.Sim.*, **42**, pp. 1317–1329.

Kapfer, S. C. and Krauth, W.(2013). Sampling from a polytope and hard-disk Monte Carlo, *J. Phys.: Conf. Ser.* **454**, 012031.

Kapfer, S. C. and Krauth, W.(2014). Two-Dimensional Melting: From Liquid-Hexatic Coexistence to Continuous Transitions, *Phys. Rev. Lett.* **114**, 035702.

Kawasaki, K.(1971). Non-hydrodynamical behavior of two-dimensional fluids, *Phys. Lett. A* **34**, pp. 12–13.

Kosterlitz, J. and Thouless, D. J.(1973). Ordering, metastability and phase transitions in two-dimensional systems, *J. Phys. C* **6**, pp. 1181–1203.

Krauth W. (2006). *Statistical Mechanics: Algorithms and Computations*, (Oxford University Press, Oxford).

Kubo, R., Toda, M. and Hashitume, N.(1991). *Statistical Physics II*, (Springer, Berlin, 1991) Chap.4.

Ladd, A. J. C., Litovitz, T. A. and Montrose, C. J.(1979). Molecular dynamics simulation o depolarized light scattering from argon at various fluid densities, *J. Chem. Phys.* **71**, pp. 4242–4248.

Ladd, A. J. C. and Alder, B. J.(1989). Decay of angular correlations in hard-sphere fluids, *J. Stat. Phys.* **57**, pp. 473–482.

Lubachevsky, B. D.(1991). How to simulate billards and similar systems, *J. Comput. Phys.* **94**, pp. 255–283.

Marín, M., Risso, D. and Cordero, P.(1993). Efficient algorithms for many-body hard particle molecular dynamics, *J. Comput. Phys.* **109**, pp. 306–317.

Marín, M. and Cordero, P.(1995). An empirical assessment of priority queues in event-driven molecular dynamics simulation, *Comput. Phys. Commun.* **92**, pp. 214–224.

Mayer, J. E. and Wood, W. W.(1965). Interfacial Tension Effects in Finite, Periodic, Two]Dimensional Systems, *J. Chem. Phys.* **42**, pp. 4268–4274.

Meel, J. A. van, Fillion, F., Valeriani, C. and Frenkel, D.(2012). A parameter-free, solid-angle based, nearest-neighbor algorithm, *J. Chem. Phys.* **136**, 234107.

Michel, M., Kapfer S. C. and Krauth, W.(2014). Generalized event-chain Monte Carlo: Constructing rejection-free global-balance algorithms from infinitesimal steps, *J. Chem. Phys.* **140**, 054116.

Michel, M., Mayer, J. and Krauth, W.(2015). Event-chain Monte Carlo for classical continuous spin models, *Euro. Phys. Lett.* **112**, 20003.

Metropolis, N., Rosenbluth, A. W., Rosenbluth, M. N., Teller, A. H., and Teller, E. (1953). Equation of State Calculations by Fast Computing Machines, *J. Chem. Phys.* **21**, pp. 1087–1092.

Nakano, H.(1993). *Linear Response Theory: a histrical perspective, Int. J. Mod. Phys. B* **7**, pp. 2397–2467.

Nelson, D. R. and Halperin, B. I.(1979). Dislocation-mediated melting in two dimensions, *Phys. Rev. B* **19**, pp. 2457–2484.

Nelson, D. R.(2002). *Defects and Geometry in Condensed Matter Physics*, (Cambridge University Press, Cambridge, England).

Nishikawa, Y., Michel, M., Krauth, W. and Hukushima, K.(2015). Event-chain algorithm for the Heisenberg model: Evidence for z $\sim$ 1 dynamic scaling, *Phys. Rev. E* **92**, 063306.

Nosenko, V., Zhdanov, S. K., Ivlev, A. V., Knapek, C. A. and Morfill, G. E.(2009). 2D Melting of Plasma Crystals: Equilibrium and Nonequilibrium Regimes, *Phys. Rev. Lett.* **103**, 015001.

Paul, G.(2007). A Complexity O(1) priority queue for event driven molecular dynamics simulations, *J. Comput. Phys.* **221**, pp. 615–625.

Pomeau, Y. and Résibois, P.(1975). Time dependent correlation functions and mode-mode coupling theories, *Phys. Rep.* **19**, pp. 63–139.

Qi, W, Wang, Z, Han, Y. and Chen, Y.(2010). Melting in two-dimensional Yukawa systems: A Brownian dynamics simulation, *J. Chem. Phys.* **133**, 234508.

Qi, W., Gantapara, A. P. and Dijkstra, M.(2014). Two-stage melting induced by dislocations and grain boundaries in monolayers of hard spheres, *Soft Matt.* **10**, pp. 5449–5457.

Rapaport, D. C.(1980). The event scheduling problem in molecular dynamics simulation, *J. Comput. Phys.* **34**, pp. 184–201.

Rapaport, D. C. (2004). *The Art of Molecular Dynamics Simulation*, (Cambridge University Press, Cambridge, 2nd Edition)

Saito, Y.(1982). Melting of Dislocation Vector Systems in Two Dimensions, *Phys. Rev. Lett.* **48**, pp.1114-1117.

Steinhardt, P. J., Nelson, D. R. and Ronchetti, M.(1983). Bond-orientational order in liquids and glasses, *Phys. Rev. B* **28**, pp. 784–805.

Stillinger, F. H.(2015), *Energy Landscapes, Inherent Structures, and Condensed-Matter Phenomena*, (Princeton University Press, Princeton)

Tobochnik, J. and Chester, G. V.(1982). Monte Carlo study of melting in two dimensions, *Phys. Rev. B* **25**, pp. 6778–6798.

Wainwright, T. E., Alder, B. J. and Gass, D. M.(1971). Decay of Time Correlations in Two Dimensions, *Phys. Rev. A* **4**, pp. 233–237.

Wood W. W. and Jacobson, J. D. (1957). Preliminary Results from a Recalculation of the Monte Carlo Equation of State of Hard Spheres, *J. Chem. Phys.* **27**, pp. 1207–1208.

Wood, W. W. (1986). Early History of Computer Simulations in Statistical Mechanics, in *Proceedings of the International School of Physics* $\ll$ *Erinco Fermi* $\gg$ *Course XCVII, Varenna on Lake Como, 1985*, G. Ciccotti and W. G. Hoover (eds.), (North-Holland, Amsterdam) pp. 3–14.

Young, A. P.(1979). Melting and the vector Coulomb gas in two dimensions, *Phys. Rev. B* **19**, pp. 1855–1866.

Chapter 7

Reflections on the Glass Transition

Jean-Marc Bomont,[a] Jean-Pierre Hansen,[b,c] and Giorgio Pastore[d]

[a] Université de Lorraine, LCP-A2MC, EA 3469, 1 Bd. François Arago, Metz F-57078, France
[b] Université Pierre et Marie Curie, UMR 8234 PHENIX, Paris, France and
[c] Department of Chemistry, University of Cambridge, Cambridge CB2 1EW, UK
[d] Università di Trieste, Dipartimento di Fisica, strada Costiera 11, 34151 Grignano (Trieste), Italy

Abstract

We present a brief overview of two key aspects of the structural glass transition (GT). The first part analyzes the structural slowing down in supercooled liquids, culminating in the kinetic GT predicted by mode-coupling theory (MCT); the theoretical predictions are confronted with the results of Molecular Dynamics simulations. The second part of the review examines the still controversial "random first order transition" (RFOT) from a quenched liquid to an "ideal glass" expected to occur at much lower temperatures, where equilibration times diverge, making the RFOT inaccessible to experiment and simulation. A "pedestrian" but robust version of the replica formalism (involving two weakly coupled replicae) is combined with integral equations for the pair structure of such a symmetric binary system, and leads to the prediction of a thermodynamic, weakly first-order RFOT both for the "soft sphere" and Lennard-Jones models, via qualitatively identical scenarios.

7.1 Introduction

When a liquid is quenched below the freezing temperature T_f by rapid cooling at constant pressure or density, for example, it undergoes a number of thermodynamic and dynamical transformations towards an amorphous solid or glass. Provided the quench rate is sufficiently fast to bypass crystal nucleation, the most common scenario involves the following transitions (in order of decreasing temperature):

- A "kinetic glass transition" (kinetic GT) characterized by dramatic structural slowing down around a temperature T_{kin}, the definition of which will be clarified in Section 7.2.

- A "laboratory glass transition" at a temperature T_g, where relaxational time scales become comparable to experimental time scales, which are typically on the order of 10^3 sec. T_g is not an intrinsic property of the glass former, but depends on the quench rate: the slower the cooling rate, the lower T_g. The "laboratory GT" is signaled by a rapid change in slope of measured thermodynamic quantities, like the molar volume (see e.g., [Debenedetti and Stillinger (2001)]).

- Below T_g, "aging" of the system sets in, i.e., measured time-dependent correlations (like the intermediate scattering function probed by inelastic neutron scattering) are no longer time-translational invariant. Moreover, the local molecular dynamics are increasingly heterogeneous ("dynamical heterogeneity") with quiescent (or frozen) nano-scale regions alternating with regions undergoing structural rearrangements at any instant of time (for a review, see [Sillescu (1999)]).

- At the so-called Kauzmann temperature T_K, the configurational entropy (to be defined later) of the strongly supercooled liquid vanishes [Kauzmann (1948)], meaning that the accessible number of metastable thermodynamic states no longer grows exponentially with the number N of atoms.

- It has long been conjectured that a "random first order transition" (RFOT) from a deeply supercooled liquid phase to an "ideal" glass phase occurs at some critical temperature T_{cr}. On the basis of formal analogies with the mean-field prediction of certain classes of spin glasses (corresponding to infinite dimensionality), it is generally accepted that T_{cr}=T_K. This also holds for structural glasses [Kirkpatrick and Wolynes (1987a,b); Kirkpatrick *et al.* (1989)].

Despite a considerable amount of recent experimental, theoretical and simulation work and numerous debates (for extensive reviews see [Angell (1991); Berthier and Biroli (2011); Wolynes and Lubchenko (2012)]), a comprehensive theory of all static and dynamical aspects of the glass transition is still missing. Although, to the best of our knowledge, Berni Alder has not worked directly on this major open problem of Condensed Matter Science, two of his ground-breaking discoveries in the 1950s and 1960's have a direct bearing on our present understanding of the glass transition. On the one hand, his pioneering Molecular Dynamics (MD) simulations, together with the late Tom Wainwright, of hard disk and hard sphere systems demonstrated the essential entropic nature of freezing [Alder and Wainwright (1957, 1962)]. Entropy provides precisely a similar driving force for structural slowing down, and eventually for the RFOT, as may be qualitatively understood in terms of the topology of the (3N+1)-dimensional potential energy landscape [Adam and Gibbs (1965); Goldstein (1969); Stillinger (1995); Debenedetti and Stillinger (2001)].

The second major discovery by Alder and Wainwright is the existence of a slow, long-time tail of correlation functions, which decay like $t^{-d/2}$ (where d is the spatial dimensionality) [Alder and Wainwright (1967, 1970)]. Alder and Wainwright showed that in the case of the velocity autocorrelation function, the slow decay is a consequence of the coupling between the motion of a tagged particle and the hydrodynamic modes of the fluid (the hydrodynamic back-flow effect; for a review, see [Pomeau and Resibois (1975)]). This seminal observation led to the "mode coupling theory" (MCT) of the slow decay of time dependent correlation functions [Gaskell and Miller (1978); Bosse *et al.* (1978)], based on the Mori-Zwanzig projection operator formalism [Hansen and McDonald (2013)]. The MCT was subsequently extended to investigate the dynamical slowing down and "structural arrest" in supercooled liquids (for an authoritative review, see [Götze (2009)]).

In this paper, we briefly review the predictions of MCT and MD simulations for the kinetic GT (Section 7.2) before focusing on the RFOT within the replica formalism first introduced by Edwards and Anderson [Edwards and Anderson (1975)] in their seminal investigation of spin glasses (Sections 7.3 and 7.4). Some recent, unpublished results will be presented in Section 7.5, while possible trails for future work will be suggested in the conclusion.

7.2 Kinetic glass transition

Upon lowering the temperature below the freezing temperature T_f, the shear viscosity η of a liquid increases rapidly [Angell (1991)], while the self-diffusion coefficient D drops at a similar rate, as expected from the

Stokes-Einstein relation:

$$D = \frac{k_B T}{2\eta\sigma} \tag{7.1}$$

where σ is the atomic diameter, and "slip" boundary conditions are assumed [Hansen and McDonald (2013)]. Radiation scattering and dielectric relaxation experiments [Angell (1991)] signal a dramatic slowing down of the atomic dynamics (compatible with the rapid drop of D), which may be characterized by wave-number k and time-dependent correlation functions, like the intermediate scattering function [Hansen and McDonald (2013)]

$$\begin{aligned} F(k,t) &= \frac{1}{N}\langle \rho_{\mathbf{k}}(t)\rho_{-\mathbf{k}}(0)\rangle \\ &= \frac{1}{N}\sum_{i=1}^{N}\sum_{j=1}^{N}\left\langle e^{i\mathbf{k}[\mathbf{r}_i(t)-\mathbf{r}_j(0)]}\right\rangle, \end{aligned} \tag{7.2}$$

where $\mathbf{r}_i(t)$ denotes the position of atom i at time t, and brackets denote statistical averages.

The normalized correlation function $\phi(k,t) = F(k,t)/S(k)$ satisfies the following generalized Langevin equation [Hansen and McDonald (2013)]:

$$\ddot{\phi}(k,t) + \Omega_k^2\phi(k,t) + \int_0^t M(k,t-t')\dot{\phi}(k,t')dt' = 0, \tag{7.3}$$

where dots denote time derivatives, the square of the characteristic frequency is $\Omega_k^2 = k_B T k^2/mS(k)$, $S(k)$ is the static structure factor, and $M(k,t)$ is the memory function. Mode coupling theory leads to the following approximate expression for $M(k,t)$:

$$M(k,t) = \nu(k)\delta(t) + \Omega_k^2 m(k,t), \tag{7.4}$$

where the first term on the rhs accounts for short-time binary collisions, while the long-time part is given by [Götze (2009)]:

$$m(k,t) = \frac{1}{2V}\sum_{\mathbf{p}}\sum_{\mathbf{q}}\delta_{\mathbf{k},\mathbf{p}+\mathbf{q}}\mathcal{V}(\mathbf{k},\mathbf{p},\mathbf{q})\phi(p,t)\phi(q,t). \tag{7.5}$$

The vertex function $\mathcal{V}$ is entirely determined by the static structure factor $S(k)$. Substitution of (7.4) and (7.5) into (7.3) leads to a non-linear integro-differential equation for the correlation function $\phi(k,t)$. At high temperatures $\phi(k,t)$ is predicted to decay exponentially; as T is lowered into the supercooled regime and approaches a critical temperature T_{kin}, the relaxation is predicted to proceed in two, increasingly well separated steps. A first step (β-relaxation) leads to a plateau where $\phi(k,t)$ remains essentially constant at a value $0 < f(k) < 1$ over a rapidly increasing time interval. Above T_{kin}, the slow α-relaxation sets in and $\phi(k,t)$ decays to zero according to

$$\phi(k,t) = f(k)\Phi(t^*), \tag{7.6}$$

where Φ is a universal scaling function of $t^* = t/\tau_k(t)$ ("time-temperature superposition"). The scaling function is well approximated by a stretched exponential $\Phi(t^*) = \exp\{-(t^*)^{\beta}\}$, where the exponent $\beta < 1$ (not to be confused with $\beta-$relaxation!). According to MCT based on eqs. (7.3)-(7.5), an ergodic to non-ergodic kinetic transition occurs at $T = T_{kin}$, when the supercooled liquid is "frozen", i.e., $\lim_{t\to\infty}\phi(k,t) = f(k) > 0$, i.e. $\lim_{T\to T^+}\tau_k(T) = \infty$ and $\alpha-$relaxation is suppressed. In fact, at sufficiently long times, thermally activated jump processes will eventually cause ergodicity to be restored even for $T < T_{kin}$. Such effects can be accomodated within MCT by including the coupling between microscopic density fluctuations and those of particle current [Götze (2009); Das and Mazenko (1986); Götze and Sjögren (1987)]. Note that the predictions of MCT hold for the so-called "fragile" glass-formers [Angell (1991)], like metal alloys or ionic melts [Mezei *et al.* (1987); Signorini *et al.* (1989)], but not for network-forming "strong" systems like silicates.

Near the kinetic GT temperature T_{kin}, typical time scales, like the Maxwell relaxation time $\tau_M = \eta/G_\infty$ (where G_∞ is the shear modulus), are of the order of nanoseconds, which are accessible to inelastic neutron scattering experiments [Mezei *et al.* (1987)] as well as to MD simulations [Hansen and Yip (1995)].

During a quench, glass formation competes with crystal nucleation; to bypass the nucleation of simple one-component systems, unphysically large quench rates are required. In multi-component systems, however, crystal nucleation is strongly impeded due to concomitant phase separation upon crystallization. It was hence recognized early on that MD tests of MCT for simple atomic systems should be carried out on binary mixtures [Bernu *et al.* (1987); Pastore *et al.* (1988)]. The model used in the earliest simulations is a binary mixture of additive "soft spheres" interacting via the pair potentials:

$$v_{\alpha\beta}(r) = \epsilon\left(\frac{\sigma_{\alpha\beta}}{r}\right)^{12} \quad ; \quad 1 \leq \alpha, \beta \leq 2 \tag{7.7}$$

with $\sigma_{\alpha\beta} = (\sigma_\alpha + \sigma_\beta)/2$, where $\sigma_1 \equiv \sigma_{11}$ and $\sigma_2 \equiv \sigma_{22}$ are the diameters of the two species. The advantage of the model is that its equilibrium excess (non-ideal) properties depend only on two variables: the number concentration $x_1 = N_1/(N_1 + N_2) = N_1/N$ and the dimensionless coupling constant $\Gamma = \rho^*(T^*)^{-\frac{1}{4}}$ (where $\rho^* = N\sigma_1^3/V$ is the reduced number density, and $T^* = k_BT/\epsilon$ is the reduced temperature) rather than on the usual three variables x_1, ρ^* and T^*. The earlier simulations [Bernu *et al.* (1987); Pastore *et al.* (1988)] were carried out for a size ratio $\sigma_2/\sigma_1 = 1.4$ and mass ratios $m_2/m_1 = 2$ or 4, while subsequent simulations were based on the less asymmetric ratio $\sigma_2/\sigma_1 = 1.2$ [Miyagawa *et al.* (1988); Roux *et al.* (1989); Barrat *et al.* (1990)]. To characterize the kinetic GT, the MD simulations provide the coherent and incoherent intermediate scattering functions $\phi(k,t)$, $\phi^{(1)}(k,t)$ and $\phi^{(2)}(k,t)$, as well as their r-space equivalents, the van Hove functions $G(r,t)$, $G_s^{(1)}(r,t)$ and $G_s^{(2)}(r,t)$, where $G^{(1)}$ and $G^{(2)}$ are the self van Hove functions:

$$G_s^{(\alpha)}(r,t) = \frac{1}{N_\alpha}\left\langle \sum_{i=1}^{N} \delta\left[\mathbf{r} - \mathbf{r}_i^{(\alpha)}(0) + \mathbf{r}_i^{(\alpha)}(t)\right]\right\rangle, \tag{7.8}$$

while $G(r,t)$ can be divided into the two self functions $G_s^{(\alpha)}$ and three "distinct" functions

$$G_d^{(\alpha\beta)}(r,t) = \frac{1}{(N_\alpha N_\beta)^{\frac{1}{2}}}\left\langle \sum_{i=1}^{N_\alpha} \sum_{j=1}^{N_\beta}{}' \delta\left[\mathbf{r} - \mathbf{r}_i^{(\alpha)}(0) + \mathbf{r}_j^{(\beta)}(t)\right]\right\rangle, \tag{7.9}$$

where the primed sum means that the $i = j$ terms are left out if $\alpha = \beta$ (these terms correspond to the self functions $G_s^{(\alpha)}$). The MD simulations also provide the time-dependent mean-square displacements of atoms of the two species:

$$\delta r^2_{(\alpha)}(t) = \frac{1}{N_\alpha}\sum_{i=1}^{N_\alpha}\left\langle |\mathbf{r}_i^{(\alpha)}(t) - \mathbf{r}_i^{(\alpha)}(0)|^2\right\rangle, \tag{7.10}$$

the asymptotic slopes of which determine the diffusion constants D_α, as well as the off-diagonal components of the microscopic stress-tensor autocorrelation functions, which determine the shear viscosity η via the standard Green-Kubo relation [Hansen and McDonald (2013)]. The simulation data are in qualitative and semi-quantitative agreement with the predictions of MCT for the same model [Barrat and Latz (1990)], but also allow further insight into the microscopic dynamics, and in particular into the role of activated jump events which "smear" the kinetic transition near T_{kin}. Some of the key results may be summarized as follows:

a) $\phi(k,t)$, $\phi^{(1)}(k,t)$ and $\phi^{(2)}(k,t)$ exhibit the three-step relaxation scenario predicted by MCT as T is gradually lowered (i.e., Γ increases) towards

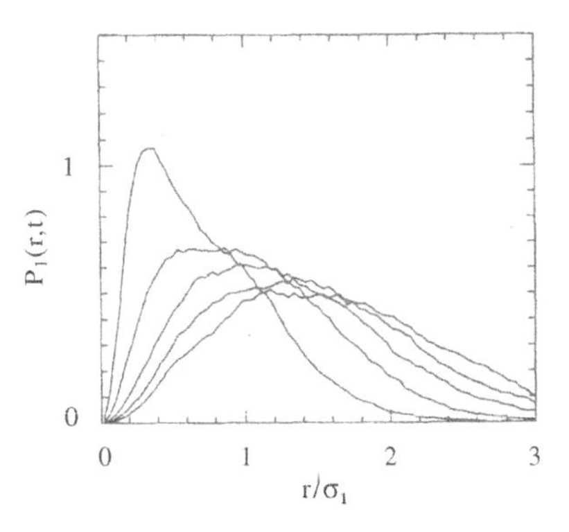

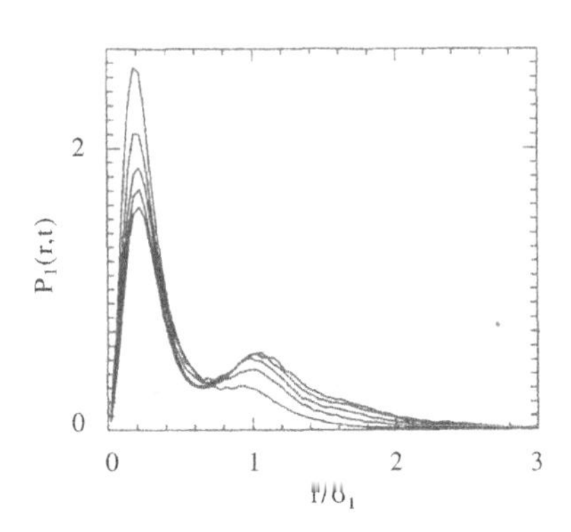

Fig. 7.1 $P_1(r,t)$ vs r for several values of t^*; left panel: $\Gamma = 1.44$; right panel: $\Gamma = 1.46$. In both panels, $t^* = 100, 200, 300, 400$, and 500 (after [Barrat *et al.* (1990)]).

T_{kin} for wavenumber k of the order of k_0 (corresponding to the main peak of $S(k)$). Note that the statistical accuracy of the $\phi^{(\alpha)}(k,t)$ is much better than that of the collective $\phi(k,t)$, since averages are taken over N_α atoms, as shown for their spatial Fourier transforms $G_s^{(\alpha)}(r,t)$ in eq. (7.8).

b) $\phi(k,t)$ and $\phi^{(\alpha)}(k,t)$ satisfy the time/temperature scaling of eq. (7.6); the relaxation time $\tau_k(T)$ varies slowly with k and the Kohlrausch exponent β tends towards 1 as k increases.

c) Since the kinetic GT is not the sharp transition at a well-defined temperature T_{kin} predicted by the "primitive" version of MCT embodied in eqn. (7.5), but is in fact "smeared" by activated jump diffusion, a working definition of T_{kin} must be adopted to characterize the rapid, but continuous transition observed in MD simulations. In practice, this is more conveniently achieved by inspecting the van Hove functions (7.8) and (7.9), rather than their Fourier transforms [Roux *et al.* (1989); Barrat *et al.* (1990)]. Consider first the self van Hove functions. The probability density for one atom of species α to diffuse a distance r at time t from its original position at t=0 is

$$P_\alpha(r,t) = 4\pi r^2 G_s^{(\alpha)}(r,t), \tag{7.11}$$

the second moment of which determines the mean square displacement (7.10).

The generic behavior of $P_\alpha(r,t)$ upon lowering T is illustrated in Fig. 7.1. At $\Gamma = 1.44$, the main peak of $P_\alpha(r,t)$ gradually moves to larger r, according to a $t^{\frac{1}{2}}$ law, as predicted by Fick's law for normal diffusion in fluids. The qualitative behavior changes dramatically around a threshold value of $\Gamma = 1.46$: the main peak is now "frozen" at a fixed value of r and its amplitude decreases slowly with time, while a secondary peak builds up at a distance roughly equal to the mean spacing between nearest neighbors. This is a clear signature of jump diffusion. The qualitative change of $P_\alpha(r,t)$ occurs over a narrow temperature interval, thus providing a clear-cut location of the kinetic GT at $\Gamma_{kin} \simeq 1.46$ in the present case [Roux *et al.* (1989); Barrat *et al.* (1990)]. The same signature of the kinetic GT was later exploited by Sastry *et al.* [Sastry *et al.* (1998)] for binary mixtures of atoms interacting via a non-additive Lennard-Jones (LJ) potential [Kob and Andersen (1995a,b)]; their simulations, which cover a substantially longer time interval than in earlier work (reflecting the steady increase in computer power!), reveal the emergence of additional secondary peaks at larger distances.

Detailed analysis of individual trajectories shows that jump diffusion proceeds by correlated, cyclic jumps of neighbouring atoms restricted to a small sub-volume of the sample [Miyagawa *et al.* (1988)], a first microscopic manifestation of a "dynamic heterogeneity," which has since been the object of numerous theoretical and numerical investigations (see e.g., [Glotzer (2000); Andersen (2005)]).

The cross-over from collective ("hydrodynamic") to jump diffusion is also mirrored in the breakdown of the Stokes-Einstein relation (7.1) around and below T_{kin} [Bernu *et al.* (1987); Pastore *et al.* (1988); Barrat *et al.* (1990)], when the D_α, measured from the asymptotic slope of the mean-square displacement (7.10), significantly exceed the prediction of (7.1); this excess illustrates the predominance of jump diffusion over the "hydrodynamic" diffusion at the basis of the Stokes-Einstein relation.

d) Structural slowing down near T_{kin} is unmistakingly diagnosed by the sudden change in the relaxation pattern of the "distinct" van Hove functions (7.9) illustrated in Fig. 7.2.

At $\Gamma = 1.44$, (i.e., above T_{kin}), $G_d^{11}(r,t)$ is seen to relax rapidly from its initial value $G_d^{11}(r,t=0) = g_{11}(r)$ to its fully decorrelated final value $\lim_{t\to\infty} G_d^{11}(r,t) = 1$. On the contrary, at the slightly lower temperature $\Gamma = 1.46$, $G_d^{11}(r,t)$ relaxes first towards an intermediate structure, which remains "frozen" over a time interval of the order of hundreds of Einstein periods ($\tau_0 \simeq (m_1\sigma_1^3/\epsilon)^{\frac{1}{2}}$), before eventually slowly relaxing towards 1. The initial decay may be associated with β-relaxation, the long-lived "frozen" structure is the r-space equivalent of the "non-ergodicity" plateau in $\phi(k,t)$, while the final decay at very long times corresponds to α-relaxation. The "life-time" of the intermediate frozen structure increases dramatically as the

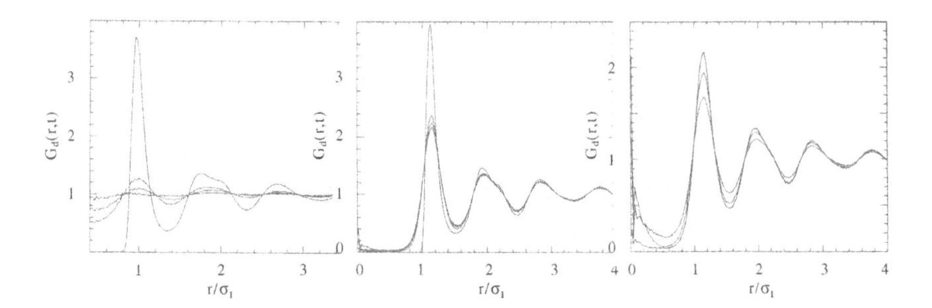

Fig. 7.2. $G_d(r,t)$ vs. r for several values of t^*; left panel: $\Gamma = 1.44$ and $t^* = 0, 60, 120$ and 240; central panel: $\Gamma = 1.46$ and $t^* = 0, 18, 36, 54$, and 72; right panel: $\Gamma = 1.46$ and $t^* = 72, 240$, and 480 (after [Roux *et al.* (1989)]).

temperature is lowered further, allowing a rather clear-cut determination of the kinetic GT temperature T_{kin} at $\Gamma_{kin} \simeq 1.46$, close to the value estimated from the cross-over behavior of $P_\alpha(r,t)$ or the break-down of the Stokes-Einstein relation.

It is important to stress that the dynamical behavior is found to be insensitive to the applied cooling rate, and that the estimates of T_{kin} (i.e., Γ_{kin}) are independent of the mass ratio m_2/m_1 [Barrat *et al.* (1990)]. The overall behavior of the dynamical slowing down of "soft sphere" mixtures has been confirmed and refined by later extensive MD simulations of binary mixtures of particles interacting via a non-additive LJ potential [Kob and Andersen (1995a,b)].

e) A kinetic GT was also detected in Brownian Dynamics simulations of concentrated suspensions of colloidal particles, where the effect of the implicit solvent is mimicked by random forces acting on the colloids [Löwen *et al.* (1991)]. The MCT scenario predicted for the slow, long-time structural relaxation is similar to that observed for deterministic Newtonian dynamics [Szamel and Löwen (1989)].

7.3 The "random first order transition" (RFOT)

As mentioned in the Introduction, mean-field theories of spin glass models, corresponding to infinite spatial dimensionality, point to the existence of an equilibrium first order transition to an "ideal glass" at a critical temperature $T_{\rm cr}$. The "ideal glass" corresponds to a thermodynamic state which is "stuck" in the lowest minimum of a suitably defined free energy "landscape" [Charbonneau *et al.* (2014)]. This RFOT concept was extended to the case of structural glasses within the framework of the density functional theory (DFT) of inhomogeneous liquids or replica theory of disordered systems [Kirkpatrick and Wolynes (1987a,b); Kirkpatrick *et al.* (1989); Berthier and Biroli (2011); Wolynes and Lubchenko (2012)]. The structural RFOT corresponds to an equilibrium transition from a supercooled liquid phase to an "ideal glass," where atoms vibrate around disordered equilibrium positions $\{\mathbf{R}_i, 1 \leq i \leq N\}$, rather than around the periodic positions of a crystal lattice. The ideal glass phase hence lacks long-range order, i.e., its one-particle density $\rho(\mathbf{r})$ is translationally invariant: $\rho(\mathbf{r}) = \rho = N/V$, just as in the liquid phase. Translational and rotational invariance also imply that the two-particle density $\rho^{(2)}(\mathbf{r}_1, \mathbf{r}_2)$ reduces to $\rho^2 g(|\mathbf{r}_1 - \mathbf{r}_2|)$; the pair distribution function $g(r)$ cannot be calculated directly by MD or MC simulations, because equilibrium cannot be reached due to divergent structural relaxation times. However, $g(r)$ is accessible, at least approximatively, by solving fluid integral equations, like the familiar hypernetted chain (HNC) equation [Hansen and McDonald (2013)]. In order to distinguish between supercooled liquid (L) and ideal glass (G) phases, an appropriate order parameter must be introduced. This is most easily achieved within the replica formalism [Franz and Parisi (1998); Mézard and Parisi (2000)]. The following discussion is restricted to the simple, but physically transparent case of two weakly coupled replicæ.

Consider a system of N atoms interacting via a pair potential $v(r)$ (replica 1) and an identical "clone" (replica 2). Atoms of opposite replicae interact via a short-ranged, purely attractive potential:

$$v_{12}(r) = -\varepsilon_{12} w(r), \tag{7.12}$$

where $\varepsilon_{12} > 0$ is the inter-replica energy scale, and $w(r)$ is a function equal to one at full overlap ($w(r = 0) = 1$), which goes rapidly to zero when r exceeds the atomic diameter σ; the exact form of $w(r)$ turns out to be irrelevant. Since the intra-replica pair potential $v(r)$ is strongly repulsive for $r \lesssim \sigma$, an atom of replica 1 interacts with at most one atom of replica 2. Using reduced coordinates $\mathbf{x}_i^\alpha = \mathbf{r}_i^\alpha/\sigma$ ($\alpha = 1, 2$), the total potential energy of this two-replicæ system reads :

$$V_{N,N}(\{\mathbf{x}_i^1\},\{\mathbf{x}_j^2\}) = \sum_i \sum_{j>i} v\left(\left|\mathbf{x}_i^1 - \mathbf{x}_j^1\right|\right) + \sum_i \sum_{j>i} v\left(\left|\mathbf{x}_i^2 - \mathbf{x}_j^2\right|\right) + \sum_i \sum_j v_{12}\left(\left|\mathbf{x}_i^1 - \mathbf{x}_j^2\right|\right). \quad (7.13)$$

The pair structure of this symmetric binary mixture is characterized by two pair distribution functions $g_{11}(x) = g_{22}(x) \equiv g(x)$ and $g_{12}(x) \equiv g'(x)$. The overlap between the configurations $\{\mathbf{x}_i^1\}$ and $\{\mathbf{x}_j^2\}$ of the two replicæ is characterized by the following overlap function [Franz and Parisi (1998)]:

$$q_{1,2}(\{\mathbf{x}_i^1\},\{\mathbf{x}_j^2\}) = \frac{1}{N}\sum_{i=1}^{N} w\left(\left|\mathbf{x}_i^1 - \mathbf{x}_i^2\right|\right). \quad (7.14)$$

The required order parameter is the mean overlap:

$$Q = \langle q_{1,2}\rangle = 4\pi\rho^* \int_0^\infty g'(x)w(x)x^2 dx. \quad (7.15)$$

If the two replicæ are completely uncorrelated, $g'(x) = 1$ and Q reduces to its "random overlap" value Q_r. In the supercooled liquid state,

$$\lim_{\varepsilon_{12}\to 0} Q(\rho^*, T^*, \varepsilon_{12}) = Q_r, \qquad \text{(L)} \quad (7.16)$$

while in the ideal glass state, the positions $\{\mathbf{x}_i^1\}$ and $\{\mathbf{x}_j^2\}$ of the atoms of the two replicæ remain localized close to the disordered equilibrium positions $\{\mathbf{X}_i = \mathbf{R}_i/\sigma\}$, so that strong inter-replicæ correlations imply that

$$\lim_{\varepsilon_{12}\to 0} Q(\rho^*, T^*, \varepsilon_{12}) \gg Q_r \qquad \text{(G)}. \quad (7.17)$$

The RFOT is hence expected to be characterized by a discontinuous jump of the order parameter Q in the thermodynamic limit [Franz and Parisi (1998)]. A related structural signature of the RFOT is $\Psi = h'(x = 0) = g'(x = 0) - 1$ which vanishes in the supercooled liquid phase when $\varepsilon_{12} \longrightarrow 0$.

In the case of the homogeneous two-replicæ system described by integral equations, the quest for solutions with $\Psi \neq 0$, in addition to the supercooled liquid, corresponds to looking for more than one solution of the integral equations. Such a request puts a constraint on the closures which may be used in connection with RFOT. In particular, a Percus-Yevick closure ([Hansen and McDonald (2013)]) for the relation between

$g'(r)$ and $v_{12}(r)$ would admit solutions with $\Psi = 0$ only. Also, the mean spherical approximation (MSA)([Hansen and McDonald (2013)]) or other closures derivable from strictly convex generating functionals must be excluded since in such cases the uniqueness of the solutions can be proved [Pastore (1988); Fantoni and Pastore (2003)].

Since in the L phase the two replicæ are uncorrelated when $\varepsilon_{12} = 0$, the calculation of thermodynamic and structural properties can be carried out on the one-component (single replica) system. The free energy $F(\rho^*, T^*)$ divides into ideal and excess parts. Since in the deeply supercooled L phase, atoms are trapped over long periods (of the order of the Maxwell relaxation time τ_M) in nearest neighbor cages, it is natural to split F into configurational (F_c) and vibrational (F_v) components, where F_v is determined by the mean vibration frequency ω_0 in a disordered medium; the latter is determined by the mean square force acting on an atom [Hansen and McDonald (2013)]:

$$\omega_0^2 = \frac{\rho}{3m} \int \nabla^2 v(r) g(r) d\mathbf{r} \tag{7.18}$$

where m is the mass of an atom. Assuming harmonic oscillations ("Einstein" model), F_v is given by

$$f_v = -\ln \left[\frac{1}{\Lambda^3} \left(\frac{2\pi k_B T}{m\omega_0^2} \right)^{3/2} \right], \tag{7.19}$$

where $f = F/Nk_BT$, and Λ is the de Broglie thermal wavelength.

Knowledge of f_{ex} allows then the calculation of the configurational free energy via:

$$\begin{aligned} f_c(\rho^*, T^*) = & \quad f_{ex}(\rho^*, T^*) + f_{id}(\rho^*, T^*) - f_v(\rho^*, T^*) & (7.20) \\ = & \quad f_{ex}(\rho^*, T^*) + \ln \left[\rho^* \left(\frac{2\pi T^*}{\omega_0^{*2}} \right)^{3/2} \right] - 1, & (7.21) \end{aligned}$$

where $\omega_0^{*2} = m\omega_0^2\sigma^3/\varepsilon$. The Kauzmann temperature T_K [Kauzmann (1948)] is defined as that at which the configurational entropy $s_c = S_c/Nk_B$ vanishes, i.e.:

$$s_c = -\left(\partial f_c^* / \partial T^*\right)_{\rho^*, T^* = T_K^*} = 0, \tag{7.22}$$

where $f_c^* = F_c/N\varepsilon = T^* f_c$; at T_K^*, $f_c^*(\rho^*, T^*)$ goes through a maximum along an isochore.

On the other hand, the RFOT critical temperature $T_{cr}(\rho^*)$ will be determined by comparing the total free energies of the supercooled liquid and ideal glass phases L and G, $f_{\mathrm{L}}(\rho^*, T^*)$ and $f_{\mathrm{G}}(\rho^*, T^*)$. In the following two sections, we present results obtained via integral equations for the "soft-sphere" model introduced in Section 7.2, and for the Lennard-Jones model.

7.4 RFOT of the "soft sphere" model

We first consider the single-component version of the "soft sphere" model defined in eq. (7), i.e.,

$$v(x) = \epsilon/x^{12}. \tag{7.23}$$

The reduced excess properties of the model depend on the single dimensionless coupling constant $\Gamma = \rho^*/T^{*1/4}$. The inter-replica potential has been chosen of the form [Mézard and Parisi (1996, 1999)]:

$$v_{12}(x) = -\varepsilon_{12} w(x) = -\varepsilon_{12} \left[\frac{c^2}{x^2 + c^2} \right]^6, \tag{7.24}$$

where c is chosen such that the range of $v_{12}(x)$ is significantly shorter than the mean distance $d^* = d/\sigma \simeq \rho^{*-1/3}$ between neighbouring atoms ($c < 1$). The precise form of $w(x)$ is irrelevant, since we will eventually be interested in the limit $\varepsilon_{12} \longrightarrow 0$. For a non-zero ε_{12}, the attraction (7.24) induces pairing of atoms of opposite replicæ into diatomic "molecules" at sufficiently low temperatures. Such molecules have a finite lifetime in the supercooled L phase, but have infinite lifetime in the G phase, as illustrated by large values $Q \gg Q_r$ of the order parameter (7.15).

We have explored the RFOT for soft spheres, using the HNC, as well as the thermodynamically consistent Rogers-Young (RY) integral equation [Rogers and Young (1984)] in two recent papers [Bomont *et al.* (2014, 2015a)]. The key findings may be summarized as follows.

7.4.1 *HNC equation*

While HNC is thermodynamically inconsistent (the virial and compressibility routes lead to significantly different equations of state), this closure allows the excess chemical potentials of the L and G phases to be expressed in terms of the pair distribution functions $g(x)$ and $g'(x)$ alone:

$$\begin{aligned} \beta\mu_{ex} &= \frac{\rho^*}{2} \int \left[h(x)\gamma(x) - 2c(x) \right] d\mathbf{x} \\ &+ \frac{\rho^*}{2} \int \left[h'(x)\gamma'(x) - 2c'(x) \right] d\mathbf{x}, \end{aligned} \tag{7.25}$$

where the total and direct correlation functions $h(x) = g(x) - 1$, $c(x)$, $h'(x) = g'(x) - 1$, $c'(x)$ are related by the familiar Ornstein-Zernike equations [Hansen and McDonald (2013)], $\gamma(x) = h(x) - c(x)$ and $\gamma'(x) = h'(x) - c'(x)$ are the indirect correlation functions, while the HNC closure relations read:

$$g(x) = \exp\left[-\beta v(x) + \gamma(x)\right], and \tag{7.26}$$

$$g'(x) = \exp\left[-\beta v_{12}(x) + \gamma'(x)\right], \tag{7.27}$$

with $\beta = 1/k_{\rm B}T$.

In the supercooled L phase, the second term of the rhs of eq. (7.25) vanishes in the limit $\varepsilon_{12} \longrightarrow 0$. The excess free energy f_{ex} follows from the thermodynamic relation $f_{ex} = \beta\mu_{ex} - \Pi_{ex}$, where $\Pi_{ex} = \beta P/\rho - 1$.

The HNC equations for $g(x)$ and $g'(x)$ are solved numerically using Gillan's iterative algorithm [Gillan (1979)] for increasing values of Γ, both for $\varepsilon_{12} \equiv 0$ (to map out the supercooled L phase) and for some initial value $\varepsilon_{12} > 0$ of the inter-replica coupling. In the latter case, a discontinuous jump of the order parameters Q and Ψ is observed at some $\Gamma_{dj}(\varepsilon_{12})$. If the discontinuity survives when the limit $\varepsilon_{12} \longrightarrow 0$ is taken, a glass phase of lower free energy is reached, which can be mapped out by solving the HNC equations keeping $\varepsilon_{12} = 0$. Glass phase solutions are obtained down to a lower value Γ_0 of Γ, below which no glass-like solutions (characterized by $Q > Q_r$ and $\Psi > 0$) exist. In fact, two glass branches, G_1 and G_2, have been detected in this way. Along the G_1 branch, $Q(\Gamma)$ and $\Psi(\Gamma)$ are found to decrease with increasing temperature (i.e., decreasing Γ) down to $\Gamma = \Gamma_0$, where Q and Ψ undergo discontinuous jumps to much higher values, corresponding to the G_2 branch, of lower free energy than the G_1 branch; when T is lowered along the G_2 branch, $Q(\Gamma)$ and $\Psi(\Gamma)$ both increase rapidly giving strong support to the identification of the G_2 branch as the "ideal" glass. A conjecture concerning the nature of the G_1 branch will be put forward in the Conclusion.

The critical value $\Gamma_{\rm cr}$ of the RFOT is determined by the intersection of the free energies of the L and G_2 branches. The HNC free energy curves $f_{\rm L}(\Gamma)$ and $f_{\rm G_2}(\Gamma)$ are found to intersect at $\Gamma_{\rm cr} \simeq 1.65$ (see Table 7.1), with

Table 7.1 HNC and RY predictions for the Kauzmann and RFOT couplings of the soft-sphere system.

	$\Gamma_{\rm K}$	$\Gamma_{\rm cr}$
HNC	1.74	1.65
RY	1.91	1.66

$f_{G_2}(\Gamma) < f_L(\Gamma)$ for $\Gamma > \Gamma_{cr}$. It is instructive to note that the configurational entropy of the G_2 phase vanishes at $\Gamma = \Gamma_{cr}$, i.e., that the Kauzmann temperature T_K associated with the ideal glass phase G_2 coincides with the critical temperature of the RFOT, thus confirming that the ideal glass is characterized by a unique disordered equilibrium configuration $\{\mathbf{X}_i\}$. In fact, by plotting $f_L(\rho^*)$ and $f_{G_2}(\rho^*)$ as functions of the volume per particle along an isotherm, one finds that the RFOT is a weakly first-order transition, with a relative volume change $(v_L^* - v_{G2}^*)/v_L^*$ determined by a Maxwell double tangent construction of the order of 1% (compared to 10% at the freezing transition). The Kauzmann coupling Γ_K listed in Table 7.1 is significantly larger than Γ_{cr}, and hence the Kauzmann "catastrophe" is preempted by the RFOT. At the RFOT, in correspondence with the critical value $\Gamma_{cr} = 1.65$, the order parameters take on very large values, $Q(\Gamma_{cr}) = 0.4212$ and $\Psi(\Gamma_{cr}) = 250.4$.

Our HNC predictions for the RFOT of soft spheres differ somewhat from the conclusions of Mézard and Parisi [Mézard and Parisi (1999)] who used a more elaborate approach based on a continuously varying number of replicæ. For each value of the coupling Γ, they minimize the Morita-Hiroike HNC free energy functional of the pair distribution function [Morita and Hiroike (1961)], which provides the state of lowest free energy, but fails to detect the metastable L or G_2 states required to characterize a first order thermodynamic transition via a Maxwell double-tangent construction. They hence predict the RFOT to be a second order transition, but their estimate of the critical coupling $\Gamma_{cr} = 1.52$ is comparable to our prediction.

7.4.2 *RY equation*

The Rogers-Young closure [Rogers and Young (1984)] allows thermodynamic self-consistency to be achieved, and its predictions for the structure and thermodynamic properties of the soft-sphere and related systems are much more accurate, compared to simulation data, than the HNC counterpart over the whole fluid range. We have hence repeated our RFOT investigations using the RY closure. The disadvantage of RY is that the HNC equation (7.25) for μ_{ex} no longer applies, so that the free energies of the L and G phases must be calculated by tedious thermodynamic integrations of the internal energies, starting from a low density (small Γ) reference state. While this poses no problem in the case of the L phase, (due to the absence of a gas-liquid transition for purely repulsive interactions), more elaborate thermodynamic paths (or "protocols") are required

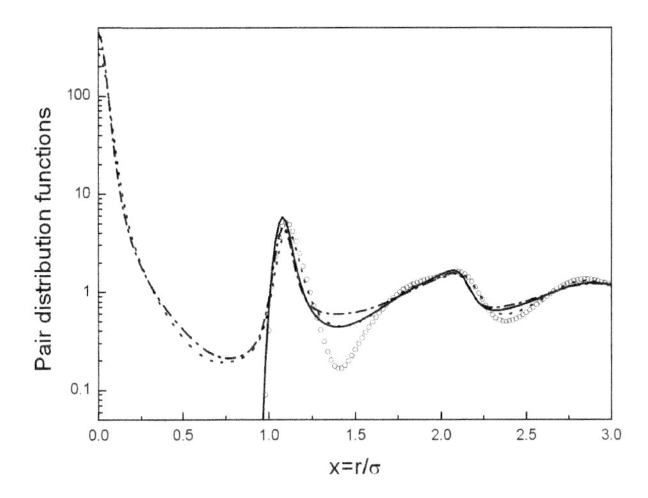

Fig. 7.3 Pair distribution functions $g(x)$ and $g'(x)$ for $\Gamma = 1.7$; HNC results: full and dash-dotted curves; RY results: dashed curve and circles.

to bypass the discontinuous jumps between the L and G branches mentioned earlier [Bomont *et al.* (2014, 2015a); Bomont and Pastore (2015)]. Apart from this technical complication, the RFOT scenario predicted by RY theory is identical to that obtained with the HNC closure, with the existence of L, G_1, and G_2 branches, and of a weakly first order transition at a coupling $\Gamma_{\rm cr}$ close to the HNC prediction (see Table 7.1). A comparison between the HNC and RY results for the L and G_2 pair distribution functions $g(x)$ and $g'(x)$ is shown in Fig. 7.3 for $\Gamma = 1.7$, illustrating the significantly enhanced structure predicted by RY.

7.5 RFOT of the Lennard-Jones model

We have extended the integral equation investigations of the RFOT to systems of atoms interacting via the Lennard-Jones pair potential $v(r) = 4\varepsilon \left[(\sigma/r)^{12} - (\sigma/r)^{6}\right]$. The structure and thermodynamic properties now depend on two independent variables, ρ^* and T^*, rather than on a single coupling constant Γ. The RFOT is hence expected to turn into a line of (weakly) first-order transitions in the density-temperature plane. The objective of the extension to a LJ system is to detect any possible qualitative changes in the scenario observed for the purely repulsive soft sphere model due to the attractive long-range interactions between atoms, which are known to greatly affect the structural slowing down [Berthier and Tarjus (2011)].

The HNC results for the LJ system [Bomont *et al.* (2015b)] confirm the RFOT scenario predicted for "soft spheres" in Section 7.4. The two-replicæ description leads to one supercooled liquid (L) branch and two glass branches G_1 and G_2, and involves discontinuous jumps in the order parameters upon varying T^* along isochores. The G_2 branch is once more tentatively identified with the "ideal glass" phase, which extends below a threshold temperature $T_0^*(\rho^*)$. The free energy curves of the L and G_2 branches intersect at a temperature $T_{\rm cr}^*(\rho^*)$ below which the "ideal glass" is the stable phase. A Maxwell double-tangent construction again leads to a small (of order 1%) relative volume change, pointing once more to a weakly first-order transition. The values of $T_{\rm K}^*$ and $T_{\rm cr}^*$ for several densities are listed in Table 7.2.

Note that systematically $T_{\rm K}^* < T_{\rm cr}^*$, so that the Kauzmann "catastrophe" is once more preempted by the RFOT. The location of the RFOT line relative to the gas-liquid-crystal phase diagram of the LJ model in the (ρ^*, T^*) plane is shown in Fig. 7.4.

Since the HNC closure becomes increasingly thermodynamically inconsistent upon increasing ρ^* or lowering T^*, we have turned to the HMSA closure [Zerah and Hansen (1986)], which generalizes the RY closure (which is applicable only to purely repulsive interactions) to the cases where the pair potential includes an attractive component. The HMSA poses the question of the optimal separation of $v(r)$ into repulsive and attractive components. For the LJ potential, we have adopted the optimized division scheme (ODS) derived by Bomont and Bretonnet [Bomont and Bretonnet (2001)], which allows numerical solutions of the HMSA equation to be found down to significantly lower temperatures than the familiar WCA scheme [Weeks *et al.* (1971); Zerah and Hansen (1986)]. The HMSA predictions for $T_{\rm K}^*$ are listed in Table 7.2 for $\rho^* = 1.0$ and $\rho^* = 1.1$; they are much lower than their HNC counterparts, a trend similar to that observed for soft spheres (see Table 7.1). The critical temperatures $T_{\rm cr}^*(\rho^*)$ at which the free energies of the L and G_2 branches cross, i.e., the HMSA location of the RFOT, are also

Table 7.2 HNC and HMSA predictions for T_K^* and $T_{\rm cr}^*$ of the LJ system.

	ρ^*	$T_{\rm K}^*$	$T_{\rm cr}^*$
HNC	1.0	0.23	0.385
HMSA	1.0	0.135	0.30
HNC	1.1	0.31	0.6
HMSA	1.1	0.18	0.59

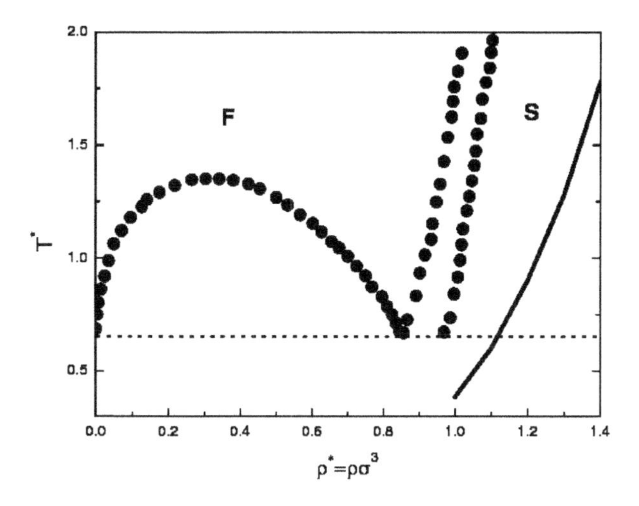

Fig. 7.4 Gas-liquid-crystal phase diagram of the LJ model in the (ρ^*, T^*) plane, together with the RFOT line predicted by HNC.

listed in Table 7.2. As expected, they are lower than the HNC predictions, while lying above the Kauzmann temperatures T_K^*. Further investigations using HMSA and other advanced integral equations [Duh and Henderson (1996)] are under way to explore higher densities.

7.6 Conclusion

While the kinetic GT in moderately supercooled liquids appears to be by now well understood, as suggested by the good agreement between the predictions of advanced mode-coupling theory and extensive computer simulations, the existence, nature, and location of the RFOT of structural glasses remains a controversial subject. We have presented some of our own recent investigations based on a two-replicæ approach and integral equations for the intra and inter-replica pair correlations within the framework of equilibrium Statistical Mechanics. Our key results for soft sphere and Lennard-Jones systems, based on the HNC, RY, and HMSA closures are summarized in Sections 7.4 and 7.5. The main conclusion is that the RFOT preempts the Kauzmann catastrophe, i.e., $T_K^*(\rho^*) < T_{cr}^*(\rho^*)$. A number of open problems remain:

a) The nature of the G_1 phase; the "unphysical" growth of the order parameters Q and Ψ as the temperature increases before a discontinuous jump towards the ideal glass (G_2) phase suggests that the two replicæ may be initially trapped in neighbouring minima of the free energy landscape, and that an increase in temperature allows the free energy barrier between these minima to be overcome, so that the two replicæ finally wind up in the lowest free energy minimum corresponding to the ideal glass.

b) Other "non-ideal" glass phases similar to G_1 may possibly be detected by switching to symmetric $m-$replicæ systems with $m > 2$ [Bomont *et al.* (2015b)]. Work along these lines is in progress.

c) It would be instructive to extend integral equations to explore the RFOT in higher spatial dimensionality d, where mean field behavior is enhanced due to the rapid increase in coordination numbers. For odd dimensionality d, the spatial Fourier transforms are expressible in terms of spherical Bessel functions which are easily handled numerically using efficient Fast Fourier Transform techniques.

d) For hard sphere systems, the glass transition goes over into a jamming transition, which has also been investigated using the replica approach [Parisi and Zamponi (2010)].

The replica theory of the RFOT has seen many recent developments, a description of which would go well beyond the scope of the present paper which is limited to a few specific aspects.

Acknowledgments

This paper is dedicated to Berni Alder whose pioneering Molecular Dynamics work with Tom Wainwright has changed the way physicists and chemists practice Statistical Mechanics forever. JPH would like to express his sincere gratitude to Berni for his constant inspiration, support, and friendship over near half a century. GP would like to thank Michele Pastore for help with figures.

Bibliography

Adam, G. and Gibbs, J. H. (1965). On the temperature dependence of cooperative relaxation properties in glassforming liquids. *Journal of Chemical Physics* **43**, 1, pp. 139–146.

Alder, B. J. and Wainwright, T. E. (1957). Phase transition for a hard sphere system. *J.Chem.Phys.* **27**, 5, pp. 1208–1209.

Alder, B. J. and Wainwright, T. E. (1962). Phase transition in elastic disks. *Physical Review* **127**, 2, pp. 359–361.

Alder, B. J. and Wainwright, T. E. (1967). Velocity autocorrelations for hard spheres. *Physical Review Letters* **18**, 23, pp. 988–990.

Alder, B. J. and Wainwright, T. E. (1970). Decay of the velocity autocorrelation function. *Physical Review A* **1**, 1, pp. 18–21.

Andersen, H. C. (2005). Molecular dynamics studies of heterogeneous dynamics and dynamic crossover in supercooled atomic liquids. *PNAS* **102**, 19, pp. 6686–6691.

Angell, C. A. (1991). Relaxation in liquids, polymers and plastic crystals, strong/fragile patterns and problems. *Journal of Non-Crystalline Solids* **131**, pp. 13–31.

Barrat, J. L. and Latz, A. (1990). Mode coupling theory for the glass transition in a simple binary mixture. *Journal of Physics Condensed Matter* **2**, 18, pp. 4289–4295.

Barrat, J. L., Roux, J. N., and Hansen, J. P. (1990). Diffusion, viscosity and structural slowing down in soft sphere alloys near the kinetic glass transition. *Chemical Physics* **149**, 1, pp. 197–208.

Bernu, B., Hansen, J. P., Hiwatary, Y., and Pastore, G. (1987). Soft-sphere model for the glass transition in binary alloys: Pair structure and self-diffusion. *Physical Review A* **36**, 10, pp. 4891–4903.

Berthier, L. and Biroli, G. (2011). Theoretical perspective on the glass transition and amorphous materials. *Reviews of Modern Physics* **83**, 2, pp. 587–646.

Berthier, L. and Tarjus, G. (2011). Testing "microscopic" theories of glass-forming liquids. *The European Physical Journal E* **34**, 9, pp. 1–10.

Bomont, J.-M. and Bretonnet, J. L. (2001). Renormalization of the indirect correlation function to extract the bridge function of simple fluids. *Journal of Chemical Physics* **114**, 9, pp. 4141–4148.

Bomont, J.-M., Hansen, J. P., and Pastore, G. (2014). An investigation of the liquid to glass transition using integral equations for the pair structure of coupled replicae. *Journal of Chemical Physics* **141**, 17, p. 174505.

Bomont, J.-M., Hansen, J. P., and Pastore, G. (2015a). Comment on 'an investigation of the liquid to glass transition using integral equations for the pair structure of coupled replicae'[J. Chem. Phys. 141, 174505 (2014)]. *Journal of Chemical Physics* **142**, 10, p. 107105.

Bomont, J.-M., Hansen, J. P., and Pastore, G. (2015b). Hypernetted-chain investigation of the random first order transition of a Lennard-Jones liquid to an ideal glass. *Physical Review E* **92**, 4, p. 042316.

Bomont, J.-M. and Pastore, G. (2015). An alternative scheme to find glass state solutions using integral equation theory for the pair structure. *Molecular Physics* **113**, pp. 2770–2775.

Bosse, J., Götze, W., and Zippelius, A. (1978). Velocity-autocorrelation spectrum of simple classical liquids. *Physical Review A* **18**, 3, pp. 1214–1221.

Charbonneau, P., Kurchan, J., Parisi, G., Urbani, P., and Zamponi, F. (2014). Fractal free energy landscapes in structural glasses. *Nature Communications* **5**, p. 4725.

Das, S. P. and Mazenko, G. F. (1986). Fluctuating nonlinear hydrodynamics and the liquid-glass transition. *Physical Review A* **34**, 3, pp. 2265–2282.

Debenedetti, P. G. and Stillinger, F. H. (2001). Supercooled liquids and the glass transition. *Nature* **410**, 6825, pp. 259–267.

Duh, D. M. and Henderson, D. (1996). Integral equation theory for LennardJones fluids: The bridge function and applications to pure fluids and mixtures, *Journal of Chemical Physics* **104**, 17, pp. 6742–6754.

Edwards, S. F. and Anderson, P. W. (1975). Theory of spin glasses. *Journal of Physics F* **5**, 5, pp. 965–974.

Fantoni, R. and Pastore, G. (2003). Generating functionals, consistency, and uniqueness in the integral equation theory of liquids. *Journal of Chemical Physics* **119**, 7, pp. 3810–3819.

Franz, S. and Parisi, G. (1998). Effective potential in glassy systems: theory and simulations. *Physica A* **261**, 3, pp. 317–339.

Gaskell, T. and Miller, S. (1978). Longitudinal modes, transverse modes and velocity correlations in liquids. I. *Journal of Physics C* **11**, 18, pp. 3749–3762.

Gillan, M. J. (1979). A new method of solving the liquid structure integral equations. *Molecular Physics* **38**, 6, pp. 1781–1794.

Glotzer, S. C. (2000). Spatially heterogeneous dynamics in liquids: insights from simulation. *Journal of Non-cristalline Solids* **274**, 1, pp. 342–355.

Goldstein, M. (1969). Viscous liquids and the glass transition: a potential energy barrier picture. *Journal of Chemical Physics* **51**, 9, pp. 3728–3739.

Götze, W. (2009). *Complex Dynamics of Glass-Forming Liquids: a Mode-Coupling Theory* (Oxford University Press, New York).

Götze, W. and Sjögren, L. (1987). The glass transition singularity. *Zeitschrift für Physik B* **65**, 4, pp. 415–427.

Hansen, J. P. and McDonald, I. R. (2013). *Theory of Simple Liquids*, 4th edn. (Elsevier, Amsterdam).

Hansen, J. P. and Yip, S. (1995). Molecular dynamics investigations of slow relaxations in supercooled liquids. *Transport Theory and Statistical Physics* **24**, 6–8, pp. 1149–1178.

Kauzmann, A. W. (1948). The nature of the glassy state and the behavior of liquids at low temperatures, *Chemical Reviews* **43**, 2, pp. 219–256.

Kirkpatrick, T. R., Thirumalai, D., and Wolynes, P. G. (1989). Scaling concepts for the dynamics of viscous liquids near an ideal glassy state. *Physical Review A* **40**, 2, pp. 1045–1054.

Kirkpatrick, T. R. and Wolynes, P. G. (1987a). Connections between some kinetic and equilibrium theories of the glass transition. *Physical Review A* **35**, 7, pp. 3072–3079.

Kirkpatrick, T. R. and Wolynes, P. G. (1987b). Stable and metastable states in mean-field Potts and structural glasses. *Physical Review B* **36**, 16, pp. 8552–8564.

Kob, W. and Andersen, H. C. (1995a). Testing mode-coupling theory for a supercooled binary Lennard-Jones mixture I: The van Hove correlation function, *Physical Review E* **51**, 5, pp. 4626–4641.

Kob, W. and Andersen, H. C. (1995b). Testing mode-coupling theory for a supercooled binary Lennard-Jones mixture. II. Intermediate scattering function and dynamic susceptibility. *Physical Review E* **52**, 4, pp. 4134–4153.

Löwen, H., Hansen, J. P., and Roux, J. N. (1991). Brownian dynamics and kinetic glass transition in colloidal suspensions. *Physical Review A* **44**, 2, pp. 1169–1181.

Mézard, M. and Parisi, G. (1996). A tentative replica study of the glass transition. *Journal of Physics A* **29**, 20, pp. 6515–6524.

Mézard, M. and Parisi, G. (1999). A first-principle computation of the thermodynamics of glasses. *Journal of Chemical Physics* **111**, 3, pp. 1076–1095.

Mézard, M. and Parisi, G. (2000). Statistical physics of structural glasses. *Journal of Physics Condensed Matter* **12**, 29, pp. 6655–6673.

Mezei, F., Knaak, W., and Farago, B. (1987). Neutron spin echo study of dynamic correlations near the liquid-glass transition, *Physical Review Letters* **58**, 6, pp. 571–574.

Miyagawa, H., Hiwatari, Y., Bernu, B., and Hansen, J. P. (1988). Molecular dynamics study of binary soft-sphere mixtures: Jump motions of atoms in the glassy state. *Journal of Chemical Physics* **88**, 6, pp. 3879–3886.

Morita, T. and Hiroike, K. (1961). A new approach to the theory of classical fluids. III general treatment of classical systems. *Progress of Theoretical Physics* **25**, 4, pp. 537–578.

Parisi, G. and Zamponi, F. (2010). Mean-field theory of hard sphere glasses and jamming. *Reviews of Modern Physics* **82**, 1, pp. 789–845.

Pastore, G. (1988). Uniqueness and the choice of the acceptable solutions of the MSA. *Molecular Physics* **63**, 4, pp. 731–741.

Pastore, G., Bernu, B., Hansen, J. P., and Hiwatari, Y. (1988). Soft-sphere model for the glass transition in binary alloys. II. Relaxation of the incoherent density-density correlation functions. *Physical Review A* **38**, 1, pp. 454–462.

Pomeau, Y. and Resibois, P. (1975). Time dependent correlation functions and mode-mode coupling theories. *Physics Reports* **19**, 2, pp. 63–139.

Rogers, F. J. and Young, D. A. (1984). New, thermodynamically consistent, integral equation for simple fluids. *Physical Review A* **30**, 2, pp. 999–1007.

Roux, J. N., Barrat, J. L., and Hansen, J. P. (1989). Dynamical diagnostics for the glass transition in soft-sphere alloys. *Journal of Physics Condensed Matter* **1**, 39, pp. 7171–7186.

Sastry, S., Debenedetti, P. G., and Stillinger, F. H. (1998). Signatures of distinct dynamical regimes in the energy landscape of a glass-forming liquid. *Nature* **393**, 6685, pp. 554–557.

Signorini, G. F., Barrat, J. L., and Klein, M. L. (1989). Structural relaxation and dynamical correlations in a molten salt near the liquidglass transition: A molecular dynamics study. *Journal of Chemical Physics* **92**, 2, pp. 1294–1303.

Sillescu, H. (1999). Heterogeneity at the glass transition: a review. *Journal of Non-Crystalline Solids* **243**, 2, pp. 81–108.

Stillinger, F. H. (1995). A topographic view of supercooled liquids and glass formation. *Science* **267**, 5206, pp. 1935–1939.

Szamel, G. and Löwen, H. (1989). Mode-coupling theory of the glass transition in colloidal systems. *Physical Review A* **44**, 12, pp. 8215–8219.

Weeks, J. D., Chandler, D., and Andersen, H. C. (1971). Role of repulsive forces in determining the equilibrium structure of simple liquids. *Journal of Chemical Physics* **54**, 12, pp. 5237–5247.

Wolynes, P. G. and Lubchenko, V. (eds.) (2012). *Structural glasses and supercooled liquids: theory,experiment and applications* (John Wiley, Hoboken).

Zerah, G. and Hansen, J. P. (1986). Self-consistent integral equations for fluid pair distribution functions: Another attempt. *Journal of Chemical Physics* **84**, 4, pp. 2336–2343.

Chapter 8

Berni Alder and Phase Transitions in Two Dimensions

J. Michael Kosterlitz

Department of Physics
Brown University
Providence, RI 02912

I do not know Berni Alder as a person, but I feel that I know him well through his seminal paper "Phase Transition in Elastic Disks" by B. J. Alder and T. E. Wainwright [1962], which was essential in motivating David Thouless and myself to think about phase transitions in two dimensional systems with a continuous symmetry. In the early 1970's, the conventional wisdom was that a crystalline solid could not exist in a two dimensional world because of the rigorous Mermin-Wagner theorem prohibiting true long range translational order at any non-zero temperature. This contradiction was settled by the theory of dislocation mediated melting to an intermediate hexatic phase followed by a second transition to the isotropic fluid at a higher temperature. This scenario, with its associated sophisticated theory, seemed to settle the controversy of two dimensional melting once and for all. However, in our elation at understanding the fundamental physics and the essential excitations of melting in 2D, we had all forgotten that the early work of Berni Alder also showed that this melting involved a weak first order transition while theory now predicted melting by two successive continuous transitions with no discontinuity in area at the critical pressure. This discrepancy could be hand waved away by arguing that Berni's system was far too small and his computers far too slow so that the areal discontinuity could be due to finite size effects or to failing to equilibrate the system. Experiments were not able to resolve the order of the transitions, but seemed to agree quantitatively with theory.

The situation was finally resolved fifty years after Alder and Wainwright's pioneering simulations on a system of 870 hard disks by a massive simulation on 2^{20} hard disks by E. P. Bernard and W. Krauth [2011]. The conclusion is that the 2D hard disk system melts by a two stage process. The crystal melts to a hexatic phase by a continuous dislocation unbinding transition following the conventional KTHNY scenario, but at a slightly higher temperature, the transition of the intermediate hexatic phase to an isotropic fluid is weakly first order with a weak areal discontinuity of similar size to that observed by Alder and Wainwright. It is remarkable that Berni Alder's 1962 simulation and the massive simulation of Bernard and Krauth in 2011 both agree that a weakly first order transition is involved in the solid/liquid transition of the 2D hard disk system.

When David Thouless and I started to think about transitions in 2D systems with a continuous symmetry, it was well known that an ordered state existed in 2D systems with a discrete symmetry, such as in the Ising and Potts models in 2D, but the rigorous Mermin Wagner theorem seemed to exclude ordering in other systems so that the existence of a crystal and a superfluid in 2D seemed to be impossible. Berni Alder's results for a 2D crystal contradicted this because, at the time, the obvious conclusion was that theory forbids a transition to an ordered state in 2D so that either the rigorous theorems or the simulations were wrong. This apparent contradiction needed to be resolved and it motivated us to construct our theory of defect mediated phase transitions [Kosterlitz and Thouless, 1973]. In 2D, this is consistent with the rigorous theorems about the absence of true long range crystalline order, but still permits a phase transition from a disordered high temperature phase to a low temperature phase with finite stiffness but no long range translational order, so that the system responds elastically to external perturbations like any conventional crystal. Berni's simulations played an essential role in our thinking while developing the dislocation theory of melting in 2D.

A few years later, our theory was developed properly by Halperin, Nelson [1978], and Young [1979], who showed that 2D melting should proceed by successive continuous transitions from a low temperature crystal melting to an intermediate hexatic fluid characterized by algebraic orientational order, but no translational order. This is then destroyed at a slightly higher temperature by another continuous transition resulting in the expected isotropic fluid phase. These results hold for a crystal described by standard elasticity theory, but of course, no crystal is completely described by elasticity theory alone and there are all sorts of extra excitations such as vacancies and interstitials which are ignored in a purely elastic theory but can affect the order of a transition. These can explain the first order nature of the second hexatic/isotropic fluid transition found by Alder and Wainwright in 1961 and Bernard and Krauth in 2011.

The significance of Berni Alder's simulations in 1962 of transitions in the 2D hard disk system cannot be overemphasized because, without these, the vast edifice of defect mediated transitions and the nature of the ordered low temperature phases of two dimensional crystals and superfluids may never have been constructed and a whole new field in condensed matter physics never invented. A whole body of physics and physicists may never have existed without Berni's contribution of a simulation well before its time. Best wishes for your 90th birthday from an admiring physicist!

Bibliography

Alder, B. J., Wainwright, T. E. (1962) Phase Transition in Elastic Disks, *Phys. Rev.* 127, pp. 359-361.

Bernard, E. P. and Krauth, W. (2011) Two-Step Melting in Two Dimensions: First-Order Liquid-Hexatic Transion, *Phys. Rev. Lett.* 107, pp. 155704.

Halperin, B. I. and Nelson, D. R. (1978) Theory of Two-Dimensional Melting, *Phys. Rev. Lett.* 41, pp. 121-124.

Kosterlitz, J. M. and Thouless, D. J. (1972) Ordering, metastability and phase transitions in two-dimensional systems, *J. Phys. C*, 6, pp. 1181- 1203.

Young, A. P. (1979) Melting and the vector Coulomb gas in two dimensions, *Phys. Rev. B*, 19, pp. 1855-1866.

Chapter 9

Molecular Dynamics of Dense Fluids: Simulation-Theory Symbiosis

Sidney Yip

Department of Nuclear Science and Engineering
Department of Materials Science and Engineering
Massachusetts Institute of Technology
Cambridge, MA 02139

Abstract

35 years ago Berni J. Alder showed the Boltzmann-Enskog kinetic theory failed to adequately account for the viscosity of fluids near solid density as determined by molecular dynamics simulation. This work, along with other notable simulation findings, provided great stimulus to the statistical mechanical studies of transport phenomena, particularly in dealing with collective effects in the time correlation functions of liquids. An extended theoretical challenge that remains partially resolved at best is the shear viscosity of supercooled liquids. How can one give a unified explanation of the so-called *fragile* and *strong* characteristic temperature behavior, with implications for the dynamics of glass transition? In this tribute on the occasion of his 90th birthday symposium, we recount a recent study where simulation, combined with heuristic (transition-state) and first principles (linear response) theories, identifies the molecular mechanisms governing glassy-state relaxation. Such an interplay between simulation and theory is progress from the early days; instead of simulation challenging theory, now simulation and theory complement each other.

9.1 Introduction

Berni J. Alder and collaborators pioneered the MD simulation of hard-sphere fluids, from liquid-solid phase stability [Alder and Wainwright, 1962] to time correlation functions and transport coefficients [Alder and Wainwright, 1970]. They showed the velocity autocorrelation function decays with a power law rather than exponentially, and explained the behavior as a hydrodynamic backflow. They also computed transport coefficients as a function of the fluid density. In this work [Alder *et al.*, 1970] the Boltzmann-Enskog kinetic theory is found to give a shear viscosity that is too low by a factor of 2 at density near that of a solid. From derivations in kinetic theory we know the Boltzmann-Enskog equation treats only the uncorrelated binary collisions. Such a description is expected to become inadequate for dense fluids where particles are close enough to each other to re-collide. Attempts have been made to incorporate the effects of correlated collisions which are expected to become important at high densities through different theoretical approaches. These include a renormalized kinetic theory formulation which incorporates correlated binary collisions [Mazneko and Yip, 1977; Yip, 1979], generalized Langevin equation with a memory function defined through projection operators [Zwanzig, 1965; Zwanzig, 2001], linear response or complex susceptibility formalism [Martin, 1968; Boon and Yip, 1991], self-consistent mode coupling theory [Götze, 2009], and equations of fluctuating hydrodynamics [Das, 2004]. All these theories are equivalent in that each involves an unknown quantity which has to be prescribed *a priori*. The unknown could appear in various forms such as a kernel of the collision integral, a space-time memory function, a self-energy propagator, a free-energy functional, or a particular mode-coupling approximation. The unknown basically describes the molecular mechanisms governing the evolution of the time correlation function of interest. In problems where the mechanisms are not known, one resorts to simplifying assumptions, or empirical models. Alternatively one could appeal to molecular simulations. In this tribute we follow the last approach to show simulation can complement theory in describing the temperature behavior of the viscosity of glasses, a phenomenon that fundamental theory on its own has not been able to fully describe.

9.2 Viscosity of supercooled liquids

Around the time of the hard-sphere simulation study, M. Goldstein proposed to analyze supercooled glass-forming liquids through a potential-energy landscape approach in which viscous flow is treated as a barrier activation process [Goldstein, 1969]. C. A. Angell later would show the experimental data on glass viscosity display distinct temperature variations in need of theoretical explanations [Angell, 1988]. Additionally a statistical mechanical framework for the glass transition phenomenon based on the energy-landscape concept of Goldstein was being formulated by F. H. Stillinger and others [Stillinger, 1988; Stillinger, 1995; Debenedetti and Stillinger, 2001]. Collectively these developments called for a molecular-level understanding analogous to the Alder challenge to kinetic theory.

The temperature variation of the shear viscosity of supercooled liquids is a longstanding problem in statistical physics. As depicted in Fig. 9.1, glass forming liquids fall into two groups, one following an Arrhenius variation while another group behave in a highly non-Arrhenius manner. The two groups are called *strong* and *fragile*, glass formers respectively A unified explanation of these two characteristic variations is of theoretical interest because it pertains to the dynamical nature of the phenomenon of glass transition (vitrification). Here we draw attention to a recent attempt that makes direct use of input from atomistic simulation to gain insight into molecular mechanisms of viscous flow by calculating the viscosity of supercooled liquids [Kushima *et al.*, 2009a; Kushima *et al.*, 2009b; Li *et al.*, 2011]. This study qualifies as a theory-simulation prediction in that no experimental information or empirical assumptions are used to account for the behavior seen in Fig. 9.1. (By simulation we do not mean the traditional molecular dynamics (MD) technique, as it is widely recognized MD has time-scale limitations rendering it ineffective for highly viscous liquids.)

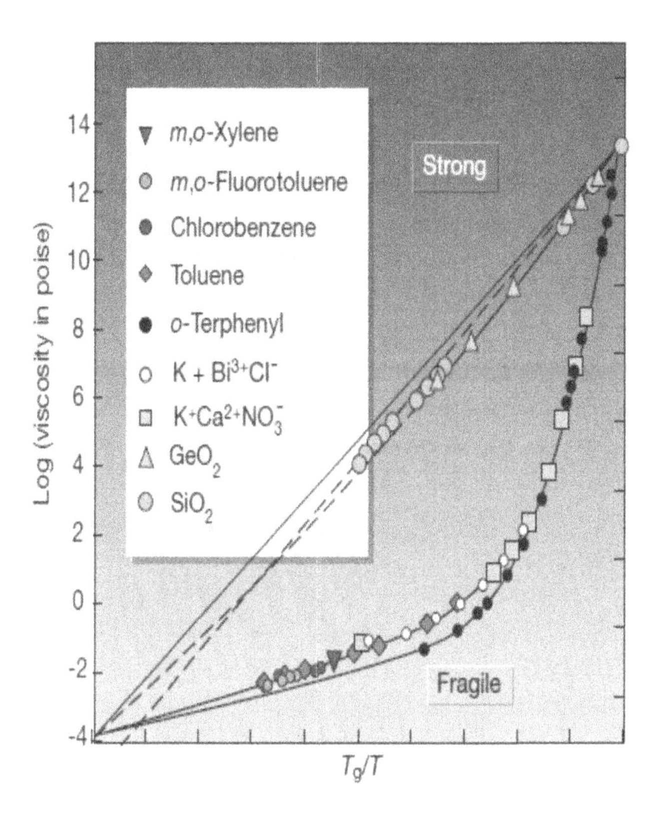

Fig. 9.1. Selected experimental data of supercooled liquids displaying characteristic *strong* and *fragile* temperature variations [Debenedetti and Stillinger, 2001]. In this tribute we explain these behavior using nonequilibrium statistical mechanics and input from atomistic simulations [Kushima *et al.*, 2009a; Kushima *et al.*, 2009b; Li *et al.*, 2011].

The basic simulation input we have in mind are the atomic trajectories the system of interacting particles follows in its evolution on the potential-energy surface. Known as transition-state pathways (TSP), they can be sampled by using a metadynamics variant of molecular simulations [Kushima *et al.*, 2009a]. A typical TSP result, shown in Fig. 9.2, is a sequence of system energies alternating between a local minimum and a saddle point. To obtain this sequence one applies an activation-relaxation algorithm which has been described in detail previously. The sampling tracks the system as it moves successively from a local minimum through a surrounding barrier (saddle) to a nearby minimum, and so on. The system does not surmount the barrier dynamically as in MD, rather the activation in time is treated according to transition-state theory. Along the TSP the atomic coordinates are determined and stored at every local minimum and saddle, so system properties of interested can be computed from this information as in any atomistic simulation. The atomic configurations thus serve as reaction coordinates in the sense of a sequence of reactions by which the system evolves on its energy surface. As such these coordinates are useful as input into a theoretical calculation either through coarse graining [Kushima *et al.*, 2009a; Kushima *et al.*, 2009b] or directly [Li *et al.*, 2011]. We illustrate the utility of TSP in both types of calculations of η(T), one based on transition-state theory, and the other on linear response formalism. In both cases the objective is to provide a molecular-level explanation of *strong* and *fragile* scaling behavior of supercooled liquids.

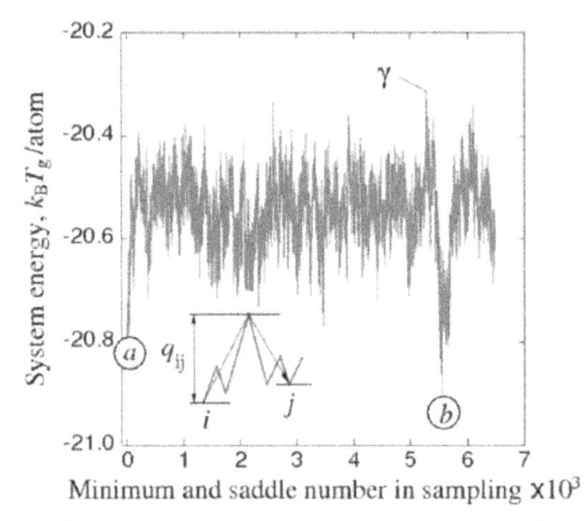

Fig. 9.2. A typical transition-state pathway (TSP) produced by metadynamics sampling [Kushima *et al.*, 2009a]. Energy is expressed in Tg which is 0.37 in reduced unit. Inset shows definition of activation barrier $q_{ij} = E_i - E_j$ as the lowest saddle point connecting energy minima i and j. This is the input from atomistic simulation for deriving an effective activation barrier in transition-state theory (see Fig. 9.3) [Kushima *et al.*, 2009a, Kushima *et al.*, 2009b], or used directly in linear response theory (Fig. 9.4) [Li *et al.*, 2011].

9.3 Understanding *strong* and *fragile* temperature scaling of supercooled liquids

In transition-state theory the viscosity is expressed through a temperature-dependent activation barrier [Kushima *et al.*, 2009a],

$$\eta(T) = \eta_0 \exp[\overline{Q}(T)/k_B T] \tag{9.1}$$

Often experimental data on η(T) are fitted to give values of $\bar{Q}(T)$, which are interpreted as activation energies when treated as constants. Also one can determine $\bar{Q}(T)$ without using Eq.(9.1), for example, by performing statistical analysis on TSP data from molecular simulations, thus calculating η(T) in a heuristic manner. This is the approach we will follow. The derivation of effective activation barriers from TSP trajectories for *fragile* [Kushima *et al.*, 2009a] and *strong* [Kushima *et al.*, 2009b] supercooled liquids has been reported previously. Although the coarse-graining analysis of a statistical series of system-level activations and relaxations to give $\bar{Q}(T)$ is instructive, for our purposes it is sufficient to consider only the results shown in Fig. 9.3.

The behavior of $\bar{Q}(T)$ as a coarse-grained activation barrier for thermally-activated processes is simple and intuitively reasonable. At the high-temperature end, triple-point conditions ($T > T_g$), the barrier is essentially temperature independent at an appropriately low value. Upon entering the "glass transition" range of supercooling, around $T \sim 2–3\ T_g$, it increases strongly, at a sharper rate for the *fragile* liquids compared to a *strong* liquid (silica in this case). One can characterize the sharpness of the activation barrier by a fragility index, $m = d\{\ln \eta(T)\}/d(T_g/T)$ at $T = T_g$. The typical experimental data shown in Fig. 9.1 would give m = 20 and 81 for silica and o-Terphenyl respectively. The limited simulation results seem to suggest the activation barrier to level off in the deeply quenched regime of $T \sim T_g$ and below. This behavior needs to be studied further. The results shown are obtained by a metadynamic sampling algorithm applied to a simple molecular model of silica (I), and a binary Lennard-Jones mixture model (II). Being derived from analysis of only simulation results using interatomic potential models they may be regarded as a first attempt at theory-simulation predictions.

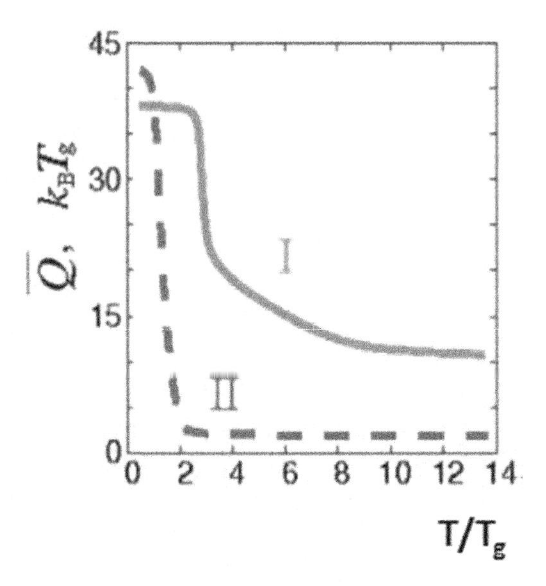

Fig. 9.3. Effective temperature-dependent activation barriers describing the viscosity of *strong* (I) [Kushima *et al.*, 2009b] and *fragile* (II) [Kushima *et al.*, 2009a] supercooled liquids.

9.3.1 *Viscosity of a* strong *glass-former*

Given Fig. 9.3, how well does one account for the experimental data in Fig. 9.1? Fig. 9.4 shows the comparison with experiment (closed circles) for silica. One sees good correspondence in the characteristic *strong* temperature scaling. Moreover, a discernable, systematic overestimate of the experimental viscosity values can be rationalized as the consequence of an upper-bound approximation in the $\bar{Q}(T)$ analysis [Kushima *et al.*, 2009b]. Direct experimental validation is significant for any theory and simulation when the data extend over 10 orders of magnitude. The range of $10^2 - 10^{12}\, Pa \cdot s$ is well beyond the reach of traditional MD simulations. The glass transition temperature for dynamical properties, usually defined as $\eta(T_g) = 10^{12}\, Pa \cdot s$, is 1446 K in this case [Angell, 1988], whereas the calculation gives 1580 K.

Besides experimental test, Fig. 9.4 also shows a comparison with two MD simulations in the high-temperature regime (where MD can be expected to be valid). The results of a comprehensive study (crosses) overlap smoothly with two independent data points obtained using a different potential model (open circles) in converging toward the high-temperature limit (as does the heuristic approach). As one goes to lower temperatures, convergence of the MD results becomes problematic due to the time-scale limitations previously mentioned. Indeed the trend of the MD data begins to deviate from the extension of the experimental data. Thus one has a gap between the available experimental and MD data. In this gap a crossover is predicted by the theory-simulation prediction (solid curve), which presents motivation for further work. An independent argument based on entropy considerations for a crossover is indicated by the arrow [Saika-Voivod *et al.*, 2001].

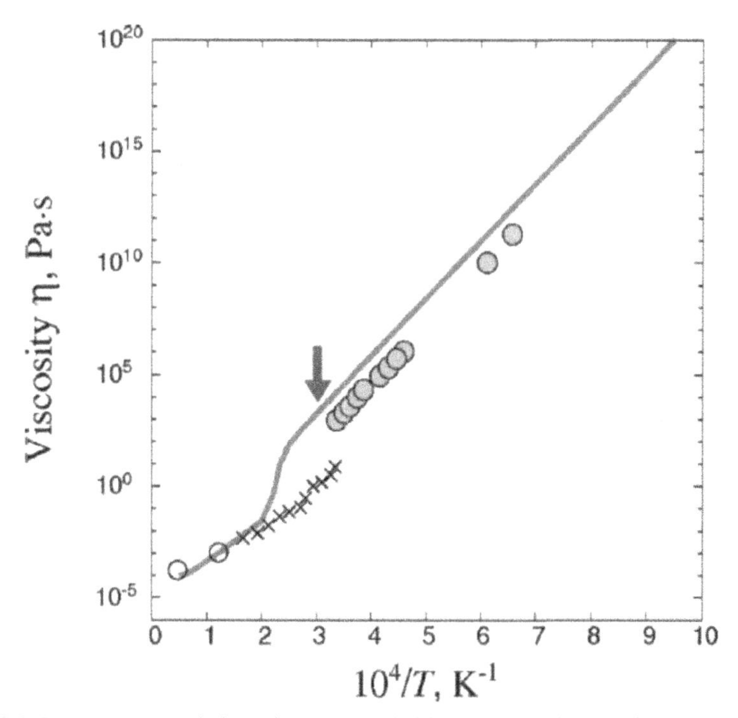

Fig. 9.4. Temperature variation of supercooled silica predicted by transition-state theory (solid line) compared to experimental data (closed circle), and MD simulation results (cross, open circle) [Kushima *et al.*, 2009b]. Arrow indicates a *strong-fragile* crossover expected from a different study [Saika-Voivod *et al.*, 2001].

9.3.2 *Viscosity of* fragile *supercooled liquids*

Two theoretical approaches have been applied to analyze *fragile* liquids. Besides using Eq. (9.1) [Kushima *et al.*, 2009a], a master equation describing a network model has been formulated in the framework of linear response theory [Li *et al.*, 2011]. As input For the transition probability connecting the network nodes, TSP data such as that shown in Fig. 9.2 were used as input directly. Fig. 9.5 shows the predictions (symbols connected by a solid line) for each of four temperature regions spanning the entire range of experimental data. Keeping in mind the BLJ potential is not designed for any single actual liquid, the agreement in this comparison of absolute values is encouraging. As with Fig. 9.4, the only parameter one needs to specify is the pre-factor ν_0 in the transition-state expression for the transition probability. This value is determined

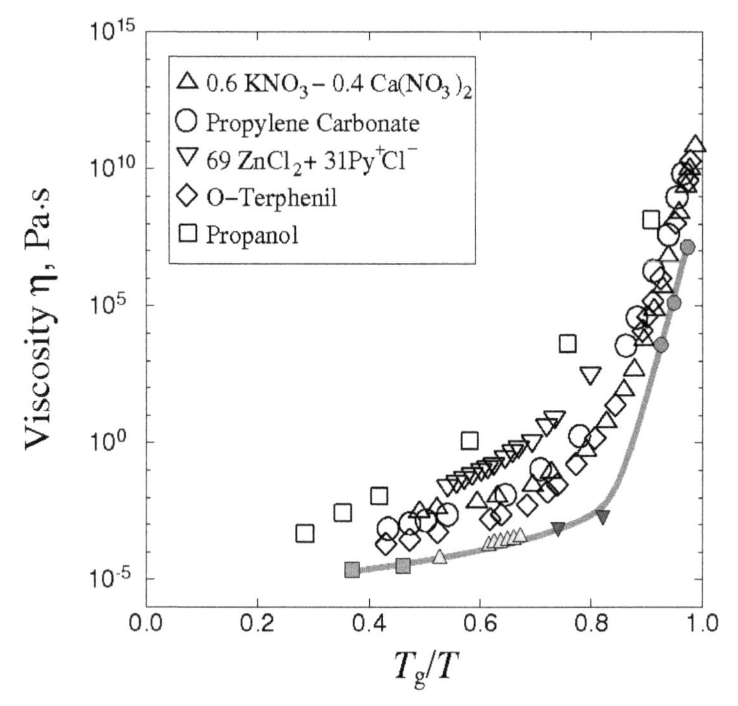

Fig. 9.5. Experimental test of linear response theory – master equation predictions (solid line) [Li *et al.*, 2011]. Symbols are experimental data of various *fragile* glass formers [Angell, 1988].

by the condition that the viscosity at high temperature approaches a value of 10^{-5} Pa·s, as seen in many experimental data. For the BLJ potential model T_g has been previously determined to be 0.37 in reduced unit [Kushima *et al.*, 2009a]. In the self-consistent mode-coupling theory which predicts an ideal glass transition, a critical temperature $T_c = 0.435$ has been estimated for the same potential [Kob and Andersen, 1995]. This gives a ratio $T_c/T_g \sim 1.3$, which is generally consistent with many observations [Angell, 1995].

A less direct test of prediction versus experiment is to scale the viscosity and invert Eq. (9.1) to give a master curve for the temperature-dependent activation barrier [Kivelson *et al.*, 1996]. This is shown in Fig. 9.6. Notice the similarity between Figs. 9.3 and 9.6. The value of T* in scaling the master-equation result is 0.63. Fig. 9.6 also shows the result of the more approximate heuristic approach which we know can be improved by including entropy corrections [Li *et al.*, 2011].

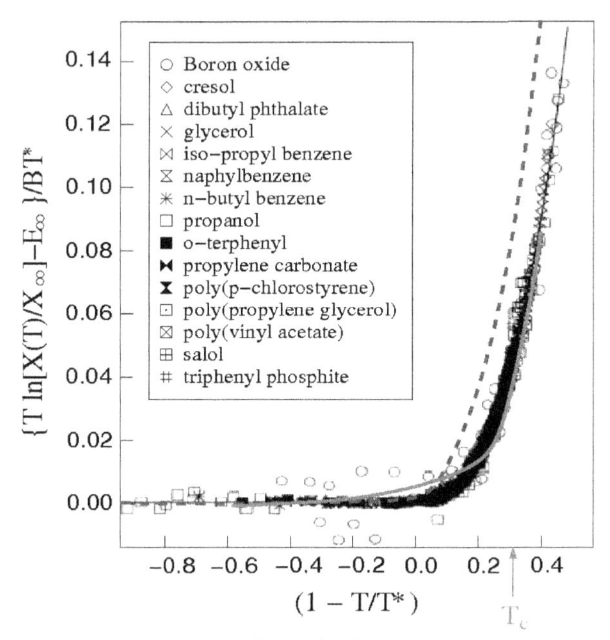

Fig. 9.6. Reduced viscosity data on fragile liquids with scaling constants, Xa, Ea, and T* in the form of an effective activation barrier [Kivelson *et al.*, 1996]. Theory-simulation results are Eq.(9.1) with $\bar{Q}(T)$ from Fig. 9.3 (case I), shown as dashed curve [Kushima, et al., 2009a], and Green-Kubo and master equation formalism as solid curve [Li *et al.*, 2011].

9.4 Towards a theory-simulation of glassy dynamics

The intention of this tribute is to draw a parallel between the understanding of transport of supercooled liquids with that of dense hard-sphere fluids in 1970. In both instances the theoretical goal was to address the dynamics of shear stress relaxation. Interestingly, what was difficult for theory in the early days, namely, the collective dynamics of correlated collisions, cage formation, and shear waves, is now difficult for MD simulations because of the challenge of sampling slow dynamics. Our discussions here suggest that to explain the distinctive temperature variations of viscosity of glasses requires a theory-simulation synergy. It is in this manner that one can begin to elucidate the underlying molecular mechanism of glassy dynamics.

At the macroscopic level the glass transition phenomenon may be viewed as a crossover from Arrhenius behavior at high temperatures to a super-Arrhenius behavior below a certain temperature, originally denoted as T_x [Angell, 1988], and more recently emphasized as a signature of supercooling [Mallemace *et al.*, 2010]. Unless one can explain the viscosity from a fundamental approach based only on molecular interactions, one does not know how to interpret the crossover. Hence the debate about the nature of this behavior, or the meaning of the glass transition, persists. The foregoing results indicate the crossover is a general manifestation of thermally activated states on a complex potential-energy surface with multiple local minima and surrounding saddle points. Given that Figs. 9.4 through 9.6 constitute a reasonable description of *strong* and *fragile* liquids from a unifying perspective, what then is the underlying molecular mechanism for the crossover? The answer seems to lie in the energy landscape becoming increasingly rough as the liquid is quenched below 2 – 3 T_g (see Fig. 9.3). Furthermore, we know from the statistical analysis that produced Fig. 9.3 [Kushima *et al.*, 2009a], this increase is partly due to the inherent structure of the supercooled liquid which is a thermodynamic property, and partly due to saddle-point activation, a kinetic property. The question of mechanism is equally relevant to the Alder challenge, stated in the Introduction. We know the viscosity underestimate by the Boltzmann-Enskog theory is almost certainly associated with the neglect of correlated collisions which becomes important at high density and would have the effect of extending the shear relaxation time and giving a higher viscosity. Accordingly, analogous to liquid vitrification by quenching one may expect a similar crossover in solid-state amorphization by compression.

The molecular mechanism of hopping across local activation barriers can take on different forms, thus enabling connections to be established between different theoretical formulations. One can imagine it to manifest in collective, dynamical process which previously have been quantified as correlated (ring) collisions, cage or cluster effects, shear wave propagation, etc. At the system-level barrier hopping process is what the self-consistent mode coupling theory would need in order to go beyond the description of ideal glass transition. (Berni and I once made a wager (an ice cream) on whether mode-coupling theory is capable of describing the dynamics of dense fluids. While I think I won the bet, I wonder if I will ever collect on it before we both forget.)

The potential energy landscape perspective is sufficiently general to apply to both *strong* and *fragile* liquids, their differences being a degree of topological roughness which one can trace to the interatomic interaction potentials and the molecular structure of the liquids. Moreover, these differences manifest in the hopping mechanisms, bond-switching (*strong*) vs. chain motions (*fragile*), that have been deciphered from the appropriate atomic configurations in the course of TSP sampling [Kushima *et al.*, 2009a; Kushima *et al.*, 2009b].

Several refinements and extensions of computational and conceptual nature are worth mentioning. The details of metadynamics simulations which we have deferred to elsewhere can surely benefit from further work with larger systems, better statistics, and more efficient algorithms. The potential energy landscape can be extended to include stress and even chemical activations in the sense of sampling free-energy surfaces. The fundamental issues of kinetics of viscous flow appear also in other rheological phenomena of interest to the materials science and statistical physics communities [Kushima *et al.*, 2011; Yip, 2016]. The combination of theory and simulation in probing molecular mechanisms at the nanoscale and connecting such insights to system behavior at the macroscale [Fan *et al.*, 2013] thus seems destined to play a significant role at the emerging mesoscale science frontier [Yip and Short, 2013; Short and Yip, 2015].

My associations with Berni has given me an appreciation of understanding complex phenomena through molecular simulation, and fond memories of his delight in challenging theorists in his unique disarming manner.

9.5 Acknowledgements

It is a pleasure to recall several years of fruitful discussions with Berni Alder, including visits to Lawrence Livermore National Laboratory, where I got to know E. Alley, Mary Ann Mansigh, and R. Pollock. I thank A. S. Argon, J. S. Langer for many discussions on glasses and viscosity of supercooled liquids, and A. Kushima, X. Lin, J. Li, J. Mauro, J. Eapen, and X.-F. Qian for collaborations. This work was supported in part by the Project on Sustainability of Kuwait's Built Environment of the MIT Center for Natural Resources and Environment and by the Basic Energy Sciences, U. S. Department of Energy award DE-SC0002633.

Bibliography

Alder, B. J., Wainwright, T. E. (1962) Phase Transition in Elastic Disks, *Phys. Rev.* 127, pp. 359-361.

Alder, B. J. and Wainwright, T. E. (1970) Decay of velocity autocorrelation function, *Phys. Rev. A* 1, pp. 18-21.

Alder, B. J., Gass, D. M., and Wainwright, T. E. (1970) Studies in Molecular Dynamics. VIII. The Transport Coefficients for a Hard-Sphere Fluid, *J. Chem. Phys.* 53, pp. 3813-3826.

Angell, C. A. (1988) Perspective on the glass transition, *J. Phys. Chem. Solids* 49, pp. 863-871.

Angell, C. A. (1995) Formation of Glasses from Liquids and Biopolymers, *Science* 267, pp. 1924-1935.

Boon, J.-P. and Yip, S. (1991) *Molecular Hydrodynamics* (Dover, New York).

Das, S. P. (2004) Mode-coupling theory and the glass transition in supercooled liquids, *Rev. Mod. Phys.* 76, pp. 785-851.

Debenedetti, P. G. and Stillinger, F. H. (2001) Supercooled liquids and the glass transition, *Nature* 410, pp. 259-267.

Fan, Y., Yildiz, B., Yip, S. (2013) Analogy between glass rheology and crystal plasticity: yielding at high strain rate, *Soft Matter* 9, pp. 9511-9514.

Goldstein, M. (1969) Viscous Liquids and the Glass Transition: A Potential Energy Barrier Picture, *J. Chem. Phys.* 51, pp. 3728-3739.

Götze, W. (2009) *Complex Dynamics of Glass-Forming Liquids* (Oxford University Press, Oxford).

Kivelson, D., Tarjus, G., Zhao, X. L., Kivelson, S. A. (1996) Fitting of viscosity: Distinguishing the temperature dependences predicted by various models of supercooled liquids, *Phys. Rev. E* 53, pp. 751-758.

Kob, W. and Andersen, H. C. (1995) Testing mode-coupling theory for a supercooled binary Lennard-Jones mixture I: The van Hove correlation function, *Phys. Rev. E* 51, pp. 4626-4641.

Kushima, A., Lin, X., Li, J., Eapen, J., Mauro, J. C., Qian, X., Diep, P., Yip, S. (2009a) Computing the viscosity of supercooled liquids, *J. Chem. Phys* 130, pp. 224504.

Kushima, A., Lin, X., Li, J., Qian, X., Eapen, J., Mauro, J. C., Diep, P., Yip, S. (2009b) Computing the viscosity of supercooled liquids. II. Silica and strong-fragile crossover behavior, *J. Chem. Phys.* 131, pp. 164505.

Kushima, A., Eapen, J., Li, J., Yip, S., Zhu, T. (2011) Time scale bridging in atomistic simulation of slow dynamics: viscous relaxation and defect activation, *Eur. Phys. J. B* 82, pp. 271-293.

Li, J., Kushima, A., Eapen, J., Lin, X., Qian, X., Mauro, J. C., Diep, P., Yip, S. (2011) Computing the Viscosity of Supercooled Liquids: Markov Network Model, *PLoS ONE* 6, pp. e17909.

Mallemace, F., Branca, C., Corsaro, C., Leone, N., Spooren, J., Chen, S.-H., Stanley, H. E. (2010) Transport properties of glass-forming liquids suggest that dynamic crossover temperature is as important as the glass transition temperature, *Proc. Nat. Acad. Sci.* 107, pp. 22457-22462.

Martin, P. C. (1968) *Measurements and Time Correlation Functions* (Gordon Breach, New York).

Mazenko, G. F. and Yip, S. (1977) *Statistical Mechanics, Part B: Time-Dependent Processes*, ed. B. J. Berni, Chapter 4 Renormalized Kinetic Theory of Dense Fluids, (Plenum Press, New York) pp. 181-231.

Saika-Voivod, S., Poole, P. H., and Sciortino, F. (2001) Fragile-to-strong transition and polymorphism in the energy landscape of liquid silica, *Nature* 412, pp. 514-517.

Short, M. P. and Yip, S. (2015) Materials aging at the mesoscale: Kinetics of thermal, stress, radiation activations, *Curr Opin Solid State Mater Sci* 19, pp. 245-252.

Stillinger, F. H. (1988) Supercooled liquids, glass transitions, and the Kauzmann paradox, *J. Chem. Phys.* 88, pp. 7818-7825.

Stillinger, F. H. (1995) A topographic view of supercooled liquids and glass formation, *Science* 267, pp. 1935-1939.

Yip, S. (1979) Renormalized Kinetic Theory of Dense Fluids, *Ann. Rev. Phys. Chem.* 30, pp. 547-577.

Yip, S. and Short, M. P. (2013) Multiscale materials modelling at the mesoscale, *Nat. Mat.* 12, pp. 774-777.

Yip, S. (2016) Understanding the viscosity of supercooled liquids and the glass transition through molecular simulations, *Molecular Simulation* 42, pp. 1330-1342.

Zwanzig, R. W. (1965) Time-Correlation Functions and Transport Coefficients in Statistical Mechanics, *Ann. Rev. Phys. Chem.* 16, pp. 67-102.

Zwanzig, R. W. (2001) *Nonequilibrium Statistical Mechanics* (Oxford Univ. Press. Oxford).

Chapter 10

Shock-Wave and Finite-Strain Equations of State at Large Expansion

Raymond Jeanloz

Earth and Planetary Science, Astronomy
Miller Institute for Basic Research in Science
University of California, Berkeley, CA 94720-4767
USA

Abstract

The empirically observed linear relationship between shock-wave velocity and particle velocity is compatible with the Eulerian finite-strain equation of state under tension as well as compression, identifying an ideal value of dynamic strength $-P_H = K_{0S}/(K_{0S}' + 1)$ and dynamic cohesive energy $E_H - E_0 = 8V_0K_{0S}/(K_{0S}' + 1)^2$ (V, K and K' are volume, bulk modulus and its pressure derivative; subscripts 0, S and H refer to zero-pressure, isentrope and Hugoniot states). The corresponding finite-strain estimate of the isentropic cohesive energy is $E_S - E_0 = 9V_0K_{0S}(2 + 2n - K_{0S}')/(2n^3)$, with strain parameter n = 2 for the Eulerian (spatial) frame of reference.

10.1 Introduction

A linear relationship between shock velocity (U_S) and particle velocity (u_p) is empirically found to describe Hugoniot equation of state measurements for a wide variety of materials

$$U = 1 + su, \tag{10.1}$$

with the zero-pressure bulk sound velocity (c_0, subscript zero indicates zero-pressure conditions) being used as normalization such that $U = U_S/c_0$ and $u = u_p/c_0$ [McQueen, et al., 1970; Marsh, 1980; Jeanloz, 1989; Trunin, et al., 2001]. The bulk sound velocity is determined by the isentropic bulk modulus (K_S) and density ($\rho = 1/V$, V is volume): $c^2 = K_S V = K_S/\rho$. Also, the slope in (1) is related to the pressure derivative of the bulk modulus, K_{0S}', by $s = (K_{0S}' + 1)/4$. Typical values of K_{0S}' are between 4 and 6, and s correspondingly lies between 1.25 and 1.75. Dynamic yielding and phase transformations cause deviations from (1), but outside the pressure-density ranges of these transitions the experimental data are well represented by the linear U–u relation over a wide range of compressions (Fig. 10.1).

The shock-wave equation of state (1) is known to be consistent with the Eulerian finite-strain (e.g., Birch-Murnaghan) equation of state under compression [Jeanloz, 1989], and the objective of the present note is to explore the consistency between these two equations of state under expansion. In principle, this amounts to exploring how well pressure-volume data can determine the cohesive energy curve for solids and liquids, including in the regime of tensile stress. Modest tensile stresses are experimentally generated by intersecting two or more release waves [e.g., Zel'dovich and Raizer, 1967, Ch XI.11]. The Eulerian finite-strain equation of state is derived from a Taylor expansion of compressional energy in terms of the spatial (Eulerian or Almansi) definition of strain.

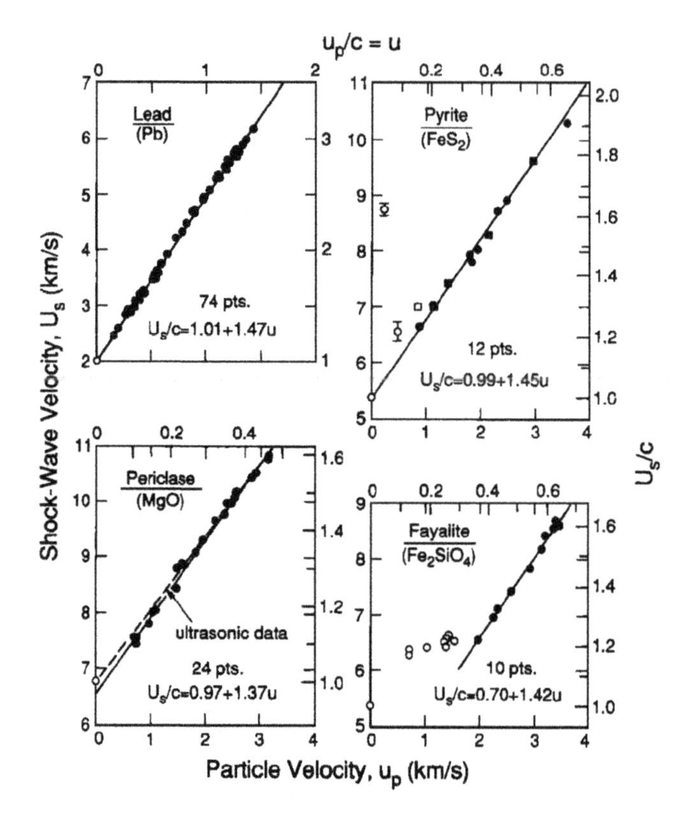

Fig. 10.1. Examples of the linear U_S–u_p relationship observed for Hugoniot data, with the ultrasonically measured value of c_0 indicated by an open symbol at $u_p = 0$ [from Jeanloz, 1989]. Fayalite undergoes a phase transition, the fit being for the high-pressure phase (closed symbols; open symbols at $0.1 < u < 0.3$ are for a partially transformed state).

10.2 Shock-wave equation of state

The Hugoniot conservation equations take on a simple form for (1) (Fig. 10.2)

$$\eta \equiv 1 - \frac{V_H}{V_0} = \frac{u}{U} = \frac{u}{1+su} \tag{10.2}$$

$$p_H \equiv \frac{P_H}{K_{0S}} = \frac{\eta}{(1-s\eta)^2} \tag{10.3}$$

$$e_H \equiv \frac{E_H - E_0}{V_0 K_{0S}} = \frac{p_H \eta}{2} = \frac{u^2}{2} \tag{10.4}$$

where E, V, P and K are internal energy, volume, pressure and bulk modulus, respectively, and subscripts H, S and 0 indicate the Hugoniot state, isentropic conditions and zero pressure [e.g., Zel'dovich and Raizer, 1967, Ch. XI; McQueen, et al., 1970]. The derivative of the Hugoniot pressure, and the shock and particle velocities are

$$\frac{dp_H}{d\eta} = \frac{(1+s\eta)}{(1-s\eta)^3} \tag{10.5}$$

$$U = \frac{1}{(1-s\eta)} \tag{10.6}$$

$$u = \frac{\eta}{(1-s\eta)} \tag{10.7}$$

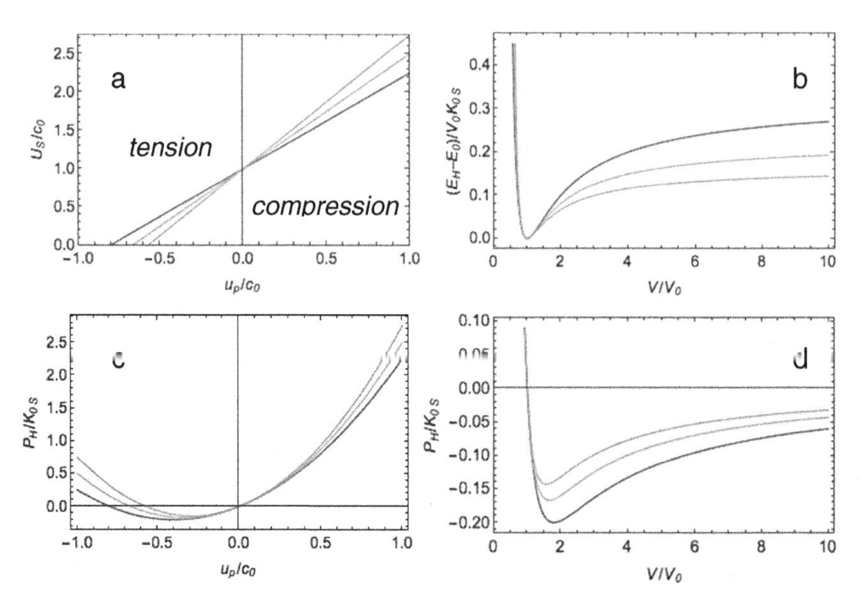

Figure 10.2. Shock-wave equation of state given by the linear U_S–u_P relation (*a*), is shown with corresponding Hugoniot energy as a function of volume (*b*), and pressure as a function of particle velocity (*c*) and volume (*d*). Both compression ($u_P > 0$, $V/V_0 < 1$, $p_H > 0$) and tension ($u_P < 0$, $V/V_0 > 1$, $p_H < 0$) are displayed for typical values of $s = 1.25$ (*blue*), 1.50 (*yellow*) and 1.75 (*green*) (equivalent to $K_{0S}' = 4$, 5 and 6 respectively). Table 10.1 summarizes key aspects of the tensile regime, $V/V_0 > 1$.

The Hugoniot energy is a minimum at $V/V_0 = 1$, rising with expansion ($V/V_0 > 1$, $u_p < 0$) as pressure becomes negative (Fig. 10.2). As can be checked using (5), the pressure achieves a minimum at a compression $\eta = -1/s$, with a value of

$$p_H\left(\eta = \frac{-1}{s}\right) = \frac{-1}{4s} = \frac{-1}{\left(K_{0S}'+1\right)}; \tag{10.8}$$

this is at a volume $V/V_0 = (1 + s)/s$, equivalent to $u = \eta/2 = -1/(2s)$ from (7). There is a change in curvature (inflection point) in the energy as a function of volume at this expansion, where the normalized shock velocity is $U = 0.5$ according to (6). Although the presence of defects causes real materials to fail under less extreme conditions, the value of

$K_{0S}/(K_{0S}' + 1)$ derived here is thought to be of the right order of magnitude for the ideal tensile strength of condensed matter [e.g., Frenkel, 1946, Ch. III.2].

With further increase in volume, the pressure rises toward zero as the energy asymptotes to a fixed value of

$$e_H\left(u = \frac{-1}{s}\right) = \frac{1}{2s^2} = \frac{8}{\left(K_{0S}'+1\right)^2}, \tag{10.9}$$

for infinite volume, a condition reached when $u = -1/s$. This state corresponds to the intercept of (1) at which the shock-wave velocity $U = 0$, and the compression η and V/V_0 are infinitely negative and positive, respectively, according to (2) (see Table 10.1).

Table 10.1. Tensile regime for shock-wave equation of state.

	Minimum Pressure*	Dynamic Cohesive Energy*
$V/V_0 = 1 - \eta$	$(1 + s)/s$	∞
$\eta = u/U$	$-1/s$	$-\infty$
$u = u_p/c_0$	$-1/(2s)$	$-1/s$
$U = U_s/c_0 = 1 + su$	0.5	0
$p_H = P_H/K_{0S}$	$-1/(4s)$	0
$e_H = (E_H - E_0)/(V_0K_{0S})$	$1/(8s^2)$	$1/(2s^2)$

*Defined by equations 10.8 and 10.9, respectively.

Thus, (9) can be thought of as the cohesive energy under dynamic expansion, as estimated from the shock-wave equation of state (1). Its magnitude of $V_0K_{0S}/(2s^2) = 8V_0K_{0S}/(K_{0S}' + 1)^2$ is 4 times larger than the energy at the inflection point that gives the minimum pressure (8), and is comparable to the thermodynamically derived cohesive energy. For example, the thermodynamic cohesive energy of Al is 12.1 ($\pm$ 0.2) MJ/kg [Kittel, 1971, p. 96; Kubaschewski and Alcock, 1979, p. 268], whereas (9) gives a value of 7.7 MJ/kg from the linear U_S-u_p relation at low pressures ($U_S = 5.33 + 1.36\ u_p$ for $u_p < 6.1$ km/s, with all velocities in km/s), and a value of 11.2 MJ/kg for the average U_S-u_p relation for Al up to 4 Gbar ($U_S = 5.94 + 1.26\ u_p$ for u_p to 350 km/s) [Trunin, et al., 2001]. Note that this is a prediction of the dynamic cohesive energy on *expansion* based on the value of K_{0S}' (or s in the linear U_S-u_p relation) obtained under *compression*.

10.3 Finite-strain equation of state

As with the shock-wave equation of state, a finite-strain expansion of energy can define an equation of state for compression along with a cohesive energy under expansion. Using a generalized definition of the finite strain measure

$$f_n = \frac{1}{n}\left[\left(\frac{V}{V_0}\right)^{\frac{-n}{3}} - 1\right], \tag{10.10}$$

the expansion of internal energy in terms f_n, normalized by V_0K_{0S} as in (4), is

$$e_S(f_n) = \frac{E_S(f_n) - E_0}{V_0K_{0S}} = 9f_n^2\left[\frac{1}{2} + \frac{a_1}{3}f_n + \frac{a_2}{4}f_n^2 + \dots\right], \tag{10.11}$$

and is differentiated to give the isentrope pressure

$$p_S(f_n) \equiv \frac{P_S(f_n)}{K_{0S}} = 3f_n(nf_n+1)^{\frac{n+3}{n}}\left[1 + a_1f_n + a_2f_n^2 + \dots\right] \tag{10.12}$$

and its derivative

$$\frac{dp_S}{df_n} = 3(nf_n+1)^{\frac{3}{n}}\left[1 + (3+2n+2a_1)f_n + 3(a_1+a_2+na_1)f_n^2 + (3+4n)a_2f_n^3 + \dots\right] \tag{10.13}$$

The strain (10) is defined as positive on compression, and values of the parameter n = 2, 1, 0 and –2 correspond to Eulerian (spatial), Biot, Hencky and Lagrangian (material) frames of reference, respectively

[Jeanloz, 1989; Holzapfel, 2000, p. 88]. Evaluating the derivatives of (12) at zero pressure (equivalent to zero strain, $f_n = 0$) gives the coefficients in terms of elastic moduli, with

$$a_1 = \frac{3}{2}\left(K_{0S}' - n - 2\right) \tag{10.14}$$

being the third-order coefficient of energy in strain (the fourth-order coefficient a_2 involves the second pressure derivative K_{0S}'').

Evidently, the strain (10) tends toward negative infinity for arbitrarily large volume expansion when $n \leq 0$, and it is only bounded when $n > 0$:

$$f_{n>0}\left(\frac{V}{V_0} \to \infty\right) = \frac{-1}{n} \tag{10.15}$$

Therefore, to leading order (setting all $a_i = 0$) the energy expansion (11) gives a finite cohesive energy only for positive values of n. The magnitude of the cohesive energy on the isentrope is, in this approximation (second order in the strain expansion of energy),

$$e_S\left(f_{n>0} = \frac{-1}{n}\right) = \frac{9}{2n^2} \tag{10.16}$$

This value is of the same order of magnitude as the dynamic cohesive energy (9) estimated from the shock-wave equation of state. For the two values to be equal it is required that

$$K_{0S}' = \frac{4n}{3} - 1 \tag{10.17}$$

which is consistent with a second-order equation of state (i.e., vanishing of the third-order coefficient a_1 defined in (14)) if $n = 9$. If the a_i coefficients in (11) are close to 1, the series in square brackets approximates ln(2), and self-consistency is achieved for $n = 9(\ln 2)^{1/2}/[4 - 3(\ln 2)^{1/2}] = 4.99$.

As (1) is a two-parameter equation of state, defined by c_0 and s (or, equivalently, by K_{0S} and K_{0S}'), (11) and (12) can be truncated at the third-order term to similarly give a two-parameter pressure–volume relation (Birch-Murnaghan equation of state). As is evident from Fig. 10.3, there is a maximum in energy (i.e., pressure becomes positive) on expansion for small n and large K_{0S}', such as $K_{0S}' = 6$ for $n = 2$ or $K_{0S}' = 5$ and 6 for $n = 1$, whereas the energy and pressure exhibit reversals upon compression for small n and small K_{0S}' (e.g., for $n = 3$ and $K_{0S}' = 4$).

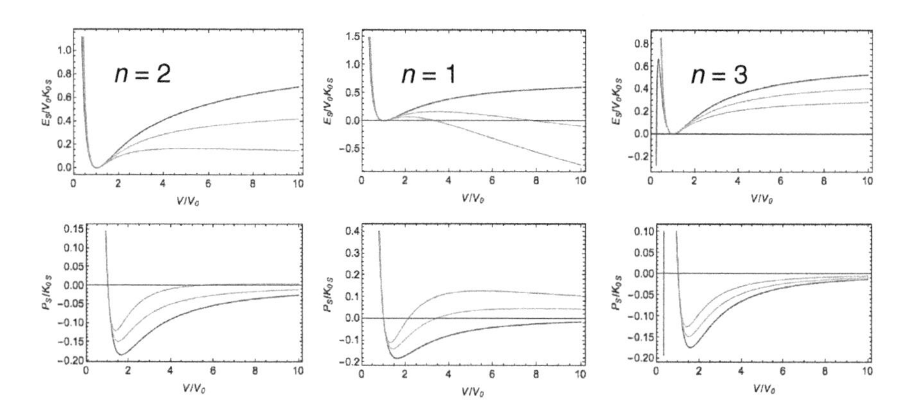

Figure 10.3. Third-order finite-strain equations of state showing the isentrope energy (*top*) and pressure (*bottom*) for strain parameter $n = 2$ (*left*), 1 (*center*) and 3 (*right*), and for values of $K_{0S}' = 4$ (*blue*), 5 (*yellow*) and 6 (*green*) (the color scheme is as in Fig. 10.2). From (14), the third-order coefficient a_1 vanishes for $K_{0S}' = 4$, 3 and 5 when $n = 2$, 1 and 3, respectively, and the minimum pressure then occurs at $V/V_0 = 1/(1 + nf_n)^{(3/n)} = [(3 + 2n)/(3 + n)]^{(3/n)} = 1.66$, 1.95 and 1.5.

The Mie–Grüneisen equation of state relates the pressure difference between Hugoniot and isentrope at a given volume to the internal energy difference (at the same volume) by way of the Grüneisen parameter γ:

$$P_H(V) - P_S(V) = \frac{\gamma}{V}\left[E_H(V) - E_S(V)\right] \tag{10.18}$$

Assuming a volume scaling

$$\gamma=\gamma_0\left(\frac{V}{V_0}\right)^q \tag{10.19}$$

the ratio of normalized pressure and energy differences obtained from (3), (12), (4) and (11) is

$$\frac{p_H(V)-p_S(V)}{\left[e_H(V)-e_S(V)\right]}=\gamma_0\left(\frac{V}{V_0}\right)^{q-1} \tag{10.20}$$

(Fig. 10.4).

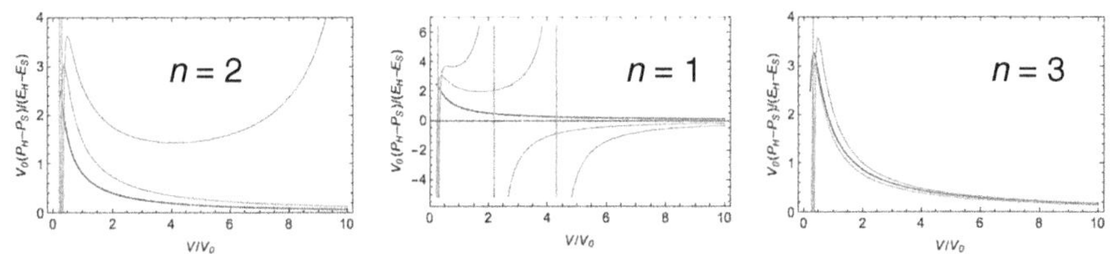

Figure 10.4. Estimated zero-pressure Grüneisen parameter (γ_0) derived from the difference between the Hugoniot and isentrope pressures and internal energies, as given by (18)-(20) for q = 1, for three values of the finite-strain parameter n defined in (10) and for K_{0S}' = 4 (*blue*), 5 (*yellow*) and 6 (*green*). The Hugoniot pressure and energy assume a linear U_S–u_p relation, (3) and (4), whereas the isentrope energy and pressure are from the finite-strain expansion, (11) and (12) with the energy expansion truncated at third-order in strain ($a_i = 0$ for $i \geq 2$).

Remarkably, (20) yields values of about the right sign and magnitude for γ_0, between 0 and 3. The curves in Fig. 10.4 are not constant, however, and rather than the usual value q = 1 assumed for the volume exponent in (19) [e.g., McQueen, et al., 1970; Jeanloz, 1989], the finite-strain and shock-wave equations of state are more consistent with each other if one assumes γ is constant (q = 0).

In detail, there are problems for small values of n (e.g., infinite and negative solutions for n = 1), but the comparison in Fig. 10.4 shows reasonable self-consistency between linear U_S–u_p and Eulerian (n = 2) finite-strain equations of state, along with the Mie-Grüneisen formulation of thermal pressure. This conclusions holds for volume expansion to $V/V_0 > 10$, as well as for compression to $V/V_0 < 0.4$. Finite-strain equations of state with larger values of n (e.g., $n = 3$) appear to fare better than those with smaller values of n.

For the cohesive energy, the third-order finite-strain equation of state gives

$$e_S\left(f_{n>0} = \frac{-1}{n}\right) = \frac{9}{2n^2}\left[1 - \frac{K_{0S}' - n - 2}{n}\right] = \frac{9}{2}\left[\frac{2n + 2 - K_{0S}'}{n^3}\right] \qquad (10.21)$$

and equating this to the dynamic cohesive energy (9) – that is, setting $e_S(f_{n>0} = -1/n) = e_H(u = -1/s)$ – yields

$$16n^3 = 9\left(2n + 2 - K_{0S}'\right)\left(K_{0S}' + 1\right)^2 \qquad (10.22)$$

as the requirement that shock-wave and finite-strain equations of state give consistent values of cohesive energy. The solution to (10.22) shows that for values of K_{0S}' between 4 and 6, a consistent strain coefficient n lies between 1 and 2.2 (Fig. 10.5). Specifically, values of $n = 1.04$ and 2.00 are compatible with $K_{0S}' = 4.00$ and 5.68, the latter being consistent with measurements on a wide variety of materials [Jeanloz, 1989].

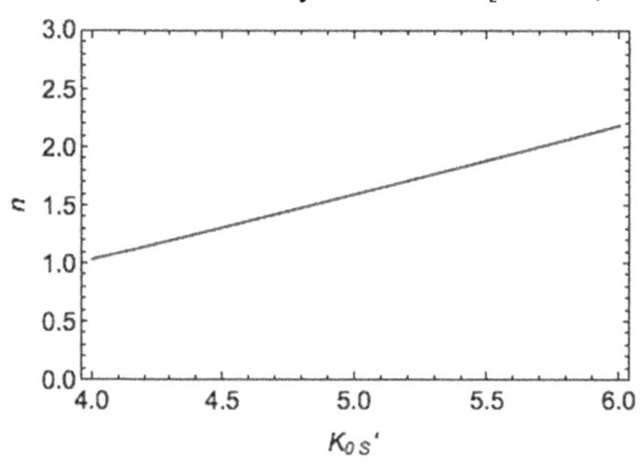

Figure 10.5. Solution of (22) showing the values of finite-strain coefficient n resulting in cohesive energies consistent with the shock-wave equation of state as a function of pressure derivative of the bulk modulus, K_{0S}' (i.e., consistency of (9) and (21)). Labeling this solution n_1, two additional solutions exist at roughly $n_2 \approx n_1 + 3.7$ and $n_3 \approx -n_1 - 0.75(K_{0S}' + 1.75)$ over this range of K_{0S}'.

10.4 Discussion

The linear U_S–u_p shock-wave equation of state gives an energy curve that increases indefinitely on compression (becoming infinite when $\eta = 1/s$), but in contrast increases on expansion toward a finite value of $1/(2s^2)$ that can be considered an estimate of the ideal dynamic cohesive energy. Similarly, the finite-strain expansion yields an energy curve that increases on compression as well as expansion. However, to leading order in strain, this isentrope energy only asymptotes to a finite value on expansion if the strain coefficient $n > 0$. Negative values of n (including the Lagrangian definition of strain) are therefore not considered.

Self-consistency between shock-wave and finite-strain equations of state is found for values close to $n = 2$, the Eulerian (spatial) definition of strain, specifically when comparing two-parameter equations of state (Fig. 10.5). Various estimates, whether for one- or two-parameter equations of state suggest consistency for values of n between about 1 and 9 (e.g., (17) and Fig. 10.5), but the general finding is that the $n = 2$ definition of strain is suitable and corrections – if needed – can be applied through higher-order terms in describing either the Hugoniot or isentrope.

In any case, there are ambiguities in the present approach that may bring detailed numerical comparison into question. Both e_H and e_S refer to adiabatic processes, yet the former is for dynamic tension and the latter is for an isentropic (that is, reversible, quasi-static) expansion. Rarefactions are generally expected to be isentropic [e.g., Zel'dovich and Raizer, 1967, Ch. XI; Jeanloz and Ahrens, 1979], and a common assumption amounts to approximating an isentrope with a linear U_S–u_p Hugoniot, at least for small strains. The observation that the differences between Hugoniot and isentrope energies and pressures, $\Delta e = e_H - e_S$ and $\Delta p = p_H - p_S$ respectively, do not vanish on expansion (or compression) means that the Hugoniot is not equated to an isentrope in the present model.

In fact, both Δe and Δp are found to be negative on expansion (i.e., the finite-strain energy and pressure are more negative than the Hugoniot energy and pressure for $V/V_0 > 1$), suggesting that dissipation is encoded into the shock-wave equation of state, yielding larger – though still negative – values of energy and pressure than the isentrope values at the same volume as estimated from the finite-strain formulation. This makes sense, to the degree that dynamic expansion may not be entirely reversible. Amazingly, the ratios of these pressure and energy differences yield quantitative estimates of the Grüneisen parameter not only of the right sign but also of the order of magnitude of expected values (Fig. 10.4).

Similarly, the cohesive energies obtained from the shock-wave and finite-strain equations of state appear to be in general agreement with expected values. There are ambiguities about comparing the dynamic and isentropic cohesive energies, as derived here, with thermodynamic estimates of cohesive energies based on heat content upon vaporization. The processes are not entirely equivalent and, just as the measured tensile strength of solids is typically less than the ideal value of "ultimate" strength, the cohesive energies discussed here may differ from those derived from thermodynamic analysis. Still, the agreement found in the present study does suggest that the shock-wave and finite-strain equations of state can be fruitfully used to describe both expansion and compression of condensed matter.

10.5 Acknowledgements

This work was supported by the Department of Energy and University of California, including UC Berkeley's Miller Institute for Basic Research in Science. I appreciate helpful discussions with Mu Li and, long ago, R. G. McQueen.

Bibliography

J. Frenkel, *Kinetic Theory of Liquids*, Oxford U. Press, Oxford, UK, 488 pp. (1946).

C. Kittel, *Introduction to Solid State Physics*, 4th Edition, Wiley, New York, NY, 766 pp. (1971).

G. A. Holzapfel, *Nonlinear Solid Mechanics*, Wiley, New York, 455 pp. (2000).

R. Jeanloz, Shock wave equation of state and finite strain theory, *J. Geophys. Res.*, 94, 5873-5886 (1989).

R. Jeanloz and T. J. Ahrens, Release adiabat measurements on minerals: The effects of viscosity, *J. Geophys. Res.*, 84, 7545-7548 (1979).

O. Kubaschewski and C. B. Alcock, *Metallurgical Thermochemistry*, 5th Edition, Pergamon Press, Oxford, UK, 449 pp. (1979).

S. P. Marsh (Ed.), *LASL Shock Hugoniot Data*, U. California Press, Berkeley, CA, 658 pp. (1980).

R. G. McQueen, S. P. Marsh, J. W. Taylor, J. N. Fritz and W. J. Carter, The equation of state of solids from shock wave studies, in *High Velocity Impact Phenomena*, edited by R. Kinslow, Academic Press, San Diego, CA, pp. 294-419 (1970).

R. F. Trunin, L. F. Gudarenko, M. V. Zhernokletov and G. V. Simakov, *Experimental Data on Shock Compression and Adiabatic Expansion of Condensed Matter*, Sarov: RFNC-VNIIEF, 446 pp. (2001).

Y. B. Zel'dovich and Y. P. Raizer, *Physics of Shock Waves and High-Temperature Hydrodynamic Phenomena*, Academic Press, San Diego, CA, 916 pp. (1967).

Chapter 11

Machine Learning and Quantum Mechanics

George Chapline

Lawrence Livermore National Laboratory, 7000 East Avenue, Livermore CA 94550

Abstract

The author has previously pointed out some similarities between self-organizing neural networks and quantum mechanics. These types of neural networks were originally conceived of as away of emulating the cognitive capabilities of the human brain. Recently extensions of these networks, collectively referred to as deep learning networks, have strengthened the connection between self-organizing neural networks and human cognitive capabilities. In this note we consider whether hardware quantum devices might be useful for emulating neural networks with human-like cognitive capabilities, or alternatively whether implementations of deep learning neural networks using conventional computers might lead to better algorithms for solving the many body Schrodinger equation.

11.1 Introduction

Among Berni Alder's many interests his lifelong pursuit of using computers to simulate many body systems has been particularly fruitful. Of course, when the many body system is a quantum system this involves well-known difficulties. When the system is at finite temperature these difficulties can to some extent be circumvented using Monte Carlo techniques, and Berni has played a leading role in this effort [Tubman, *et al.* 2012]. Unfortunately exact simulation of isolated quantum systems with more than say 100 particles has remained elusive. Not only has this impacted the ability to accurately predict the properties of materials, but this has limited the ability to simulate hardware quantum devices to devices of modest complexity. The possibility of constructing quantum computing devices which in the limit of many qubits would be able to solve problems that would be impractical to solve with existing digital computers has attracted a great deal of attention. Indeed at the present time there are considerable efforts at a number of laboratories around the world to develop hardware quantum devices for quantum information processing. In this paper we will consider the implications of recent advances in machine learning for simulating many body quantum systems with either hardware quantum devices or conventional computers.

As is well known pattern recognition algorithms for conventional classical computers have had great difficulty matching the capabilities of the mammalian brain. On the other hand great progress has recently been made as a result of the availability of large scale low cost computation which has led to the development of "deep learning neural networks" [Jordan and Mitchell, 2015]. It has now been demonstrated that not only can deep learning neural networks recognize speech and handwriting patterns about as well as a human, but can even learn abstract concepts [Lake, et al., 2015]. These successes are ultimately based on David Mumford's suggestion [Munford 1994] that the human brain makes use of Rissanen's minimum description length principle for "top-down" explanations of sensory inputs, and have narrowed the focus of machine learning research to model-based neural networks [Bishop, 2012].

It has been understood for some time that pattern recognition systems are in essence machines that utilize either preconceived probability distributions or empirically determined posterior probabilities to classify patterns [Riplet 1996]. In the ideal case where the a priori probability distribution $p(\alpha)$ for the occurrence of various classes α of feature vectors and probability densities $p(x|\alpha)$ for the distribution of data sets x within each class are known, then the best possible classification procedure would be to simply choose the class for which the posterior Bayesian probability

$$P(\alpha \mid x) = \frac{p(\alpha)p(x \mid \alpha)}{\sum_{\beta} p(\beta)p(x \mid \beta)} \tag{11.1}$$

is largest. Unfortunately in the real world one is typically faced with the situation that neither the class probabilities p(α) nor class densities p(x|α) are precisely known, so that one must rely on empirical data to estimate the conditional probabilities $P(\alpha| \text{x})$ needed to classify data sets. In practice this means that one must adopt a parametric model for the class probabilities and densities, and then use empirical data to fix the parameters of the probability model. Unfortunately determining the best possible values for the model parameters from empirical data is a computationally difficult problem, and until very recently this has limited researchers to using models of relatively modest complexity.

One popular metric for how good a particular set of model parameters is at reproducing the observed data is the maximum likelihood (ML) estimator, can be motivated by noting that the formula for the posterior probability given in (1) can be recast in the form similar to the canonical Boltzmann distribution for the populations of the energy levels of a system in equilibrium with a heat bath. In particular if one defines the "area" of a classification α to be the number of information bits required to describe the classification

$$S(\alpha) \equiv -\log p(\alpha)p(x \mid \alpha) \tag{11.2}$$

Following Mumford [2012] one could say that the aim of human-like pattern recognition is to construct models for the probabilities p(α) and p(x|α) that minimize S(α). If instead of the true probability distributions p(α) and p(x|a) one uses model probabilities p(α;θ) and p(x|α;θ) to calculate an approximate probability distribution $P(\alpha;\theta)$ for different classifications of a data set x, then (3) will no longer necessarily be satisfied and the free energy $F(\text{x},\theta)$ calculated using the distribution

$P(\alpha;\theta)$ will in general differ from the true free energy. In particular we would have that

$$F = F(\theta) - \sum_{\alpha} P(\alpha;\theta)\log[P(\alpha;\theta)/P(\alpha)] \tag{11.4}$$

The entropy-like second term in (4) is always positive and measures the difference in bits between the model distribution $P(\alpha;\theta)$ and the true distribution $P(\alpha)$, and is the basis for the ML estimator.

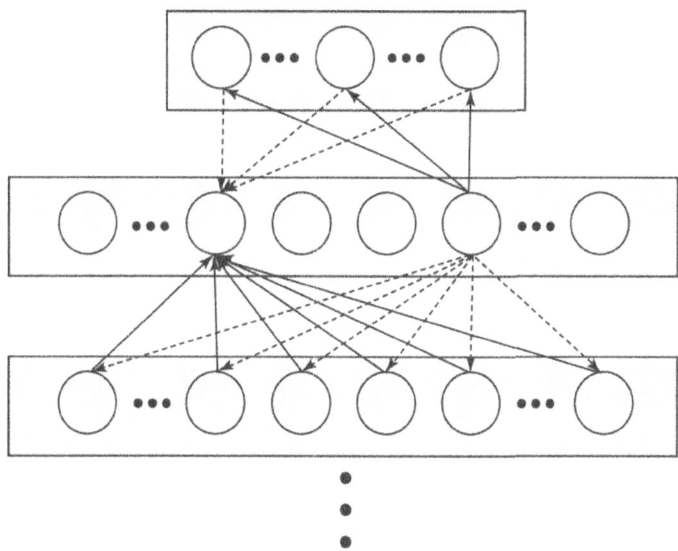

Fig. 11.1. A Helmholtz machine contains a recognition neural network intertwined with a generative model for the input data.

A promising neural network approach to modeling how the human brain actually constructs ML models for sensory data was introduced in the 1990s by Geoffrey Hinton and his students [Hinton, 1995; Dayon, 1995]. They called their network the "Helmhotlz machine" in recognition of Helmholtz's seminal idea that the mammalian brain, in contrast with more primitive brains, had the distinctive capability of being able to infer the probable cause of sensory data. The Helmhotlz machine consists of two intertwined Bayesian networks: a "top- down" network that generates models of the world a "bottom-up" network that interprets sensory data (see Fig. 11.1). In the computer models constructed by Hinton *et al.* both networks consisted of layers of units whose activation levels are either 0 or 1. The binary units in the lowest layer are used to represent environmental input data, while the remaining "hidden" units in the upper layers represent possible explanations for the input data. is the Helmholtz machine, which as originally described used a "wake-sleep" algorithm to self-consistently construct "bottom-up" and "top-down" representations of sensor data [Ripley, 2012]. During the "wake phase" the connection strengths of the top down network are modified so as to make the model activation probabilities for the units in the hidden layers align more closely with the actual activations of the these units that are observed when the bottom-up recognition network is used to interpret examples of input data that are fed to the first layer. This learning step makes the top-down layer better at constructing realistic models of the world. In the "sleep phase" the connection strengths in the bottom-up network are modified so that the activation probabilities for the units in the hidden layers of the recognition network are aligned more closely with the activities of the these units that are observed when the top-down network is used to generate "fantasies" of world. Unfortunately implementations of the Helmholtz machine's wake-sleep algorithm have been limited because Monte Carlo-like calculations are required to evaluate the joint probability distributions needed to construct "bottom-up" and "top-down" representations of the input data. Remarkably the deep learning neural networks are able to sidestep the difficulty of directly constructing conditional probability distributions by using a hierarchical structure for the generative models together with back-propagation learning.

In this paper we will discuss whether there may be some connection between human-like pattern recognition and quantum mechanics. In particular, one is tempted to interpret S(α) as the classical action for a string of spins, and the denominator of (2) as an imaginary time path integral for this spin system with a time varying Hamiltonian. In addition the way the bottom-up and top-down descriptions of input data in a Helmholtz machine relax to a ML representation for the input data resembles Feynman-Keldysh double path integral method [Feynman and

Hibbs, 1965] for describing the time evolution of a quantum system subject to quantum noise. While these hints of connections with quantum dynamics don't necessarily provide any clues as to how to improve pattern recognition using conventional computers, they do suggest that quantum devices emulating Helmholtz-like neural networks might eventually prove useful for pattern recognition [Chapline, 2004]. Alternatively, the recent advances in machine learning associated with deep learning networks prompt the question as to whether these recent advances in the capabilities of neural networks to emulate the human brain might improve the prospects for simulating quantum dynamics with conventional computers. In the following, after discussing whether hardware quantum devices might be used to emulate human cognition, we turn to the question whether software simulations of deep learning neural networks with human-like cognitive capabilities might also be useful for calculating many body quantum wave functions.

11.2 Quantum mechanics and self-organizing neural networks

Our original inspiration for the idea that self-organizing neural networks might be related to quantum mechanics was provided by a connection between a self-organizing neural net and the traveling salesman problem. A completely deterministic method for self-organizing data from an array of sensors, known as Kohonen self-organization [Kohonen, 1995], has found many applications, including providing an algorithm for solving the traveling salesman problem that compares favorably with the best known algorithms. The traveling salesman problem is solved by the simple expedient of connecting randomly placed nodes connected with each other and the cities to be visited by elastic strings, and then allowing the system to relax to the lowest energy state using gradient descent dynamics [Durban and Willshaw, 1987].

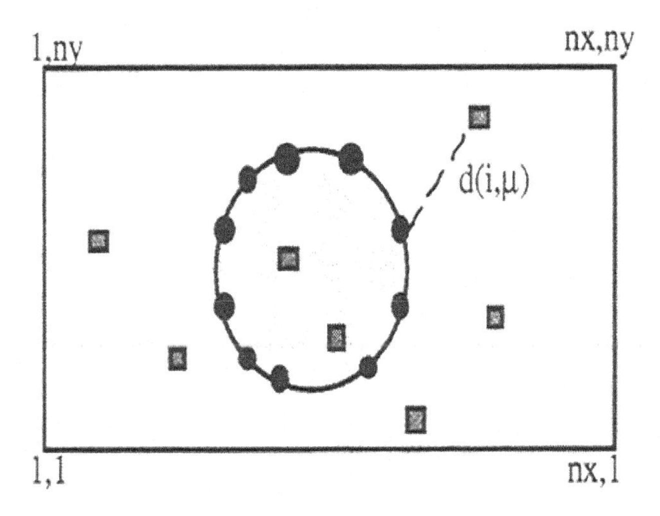

Fig. 11.2. Solution for the traveling salesman problem using self-organization.

In 2001 the author noted [Chapline, 2001] that the elastic net method for solving the traveling salesman problem is essentially equivalent to taking the classical limit of the Feynman path integral for a particle moving in a magnetic field. Translating this result to the Helmholtz machine suggests that probabilistic self-organization analog of deterministic self-organization by taking the classical limit of the path integral would be the relaxation of a double path integral for two strings of data features coupled via elastic springs. In the limit of weak coupling one can described the motion of each string of data features as the motion of a single string acted on by a stochastic classical field. It is the stochastic nature of the influence of one string on the other that provides the dissipation needed to continuously reduce the area swept out in feature space by the both the bottom-up and top-down strings, and thereby provide minimum information cost representations for the sensory data. Unfortunately, the quantum dynamics of the two coupled strings cannot be simulated in the continuum limit on a conventional computer, which is perhaps related to why simulations of the Helmholtz machine has not been widely used for data analysis. Indeed the difficulty of using conventional computers to simulate the probabilistic dynamics needed to analyze complex data provides an incentive for seeking an analogue quantum system that can be used to emulate the wake-sleep algorithm.

One example of an analog quantum system that might be used to emulate the Helmholtz machine is a quantum version [Kohmoto *et al.*, 1981] of the Ashkin-Teller (AT) statistical model for a 2-dimensional array of spins with two spins per lattice site. It was discovered by Kadanoff and Brown [1979] that if the spin couplings are carefully chosen, the energy functional for the model has a Gaussian form very similar in form to the energy cost functions that are typically used by deterministic self-organizing algorithms [Hertz *et al.,* 1991]. When the spins at each lattice site are allowed to fluctuate, these critical AT models share in common with self-organizing algorithms the crucial property that 2-dimensional of the spins will be invariant under conformal transformations of the coordinates. This means that the AT models with critical values for the spin couplings will possess string-like excitations where the spin expectation values vary smoothly with position in the array.

In order to make contact with the wake-sleep algorithm [Hinton *et al.*, 1995] one could identify the two types of spins in the AT model with the bottom-up and top-down networks of spins used in the Helmholtz machine. The cycles of the wake-sleep algorithm could then be simulated by allowing the 2-spin and 4-spin AT couplings to vary with time in such a way to both decrease the description cost for the data in the bottom-up and top-down layers of a Helmholtz machine and cause these representations to converge. In order to emulate the cycles of the wake-sleep algorithm one could periodically increase and then decrease the 4-spin coupling; causing the configuration of spin expectation values in the top-down and bottom-up layers to converge while allowing for self-organization of spin expectation values in both layers. It is noteworthy that this can be accomplished in the homogeneous AT model while preserving criticality, and therefore conformal invariance, by simultaneously changing the 2-spin coupling. What the appropriate cycling conditions might be in an inhomogeneous array of spins with

2-spin and 4-spin interactions remains to be determined. Even if the initial spin excitation values in these models are random, self-organization of spin expectation values among all the lattice points of the AT model should approximately resemble the self-organization of spins in the Helmholtz machine.

In the quantum emulation of the wake-sleep algorithm just described the input sensory data would be represented as the joint probability distribution for the bottom-up layer of quantum spins. Ideally this joint probability distribution would capture the non-classical aspects of the initial wave function representing a superposition of qubit states. The spin configurations for the top-down layer could initially chosen at random, but eventually the qubits along one edge would represent generators for the top-down models for the sensory data. In cases where the output of a sensor requires more than 1 qubit, one would need to replace each individual qubit in the quantum AT model with a cluster of qubits. The D-Wave Systems quantum processor [Harris *et al.*, 2010] may be a first step towards realizing these possibilities. Although our scheme for emulating the wake-sleep algorithm employs a quantum computational strategy that is different from the adiabatic quantum computing strategy that has been the focus of researchers using the D-Wave machine, an array of SQUID qubits similar to the D-Wave processor, both with respect to its architecture and the ability to vary coupling strengths between SQUID qubits, could in principle be used to represent the Ising spins in an AT emulation of the wake-sleep algorithm. This would require much greater quantum coherence for the SQUID array as compared with what has been achieved with the D-Wave processor, and a method for rapidly varying the couplings between the SQUID qubits. As it happens construction of superconducting qubit arrays with long coherence lifetimes and the capability of operating at radio frequencies is currently one of the main goals the quantum computing community.

11.3 Can deep learning neural networks be used to simulate many-body quantum dynamics?

The wake-sleep algorithm for the Helmholtz machine might be viewed as analogous to the "Newton-Jost problem" of finding the Hamiltonian for a multi-channel Schrodinger equation given scattering data [Newton, 1989]. Actually an explicit connection between the inverse problem for multi-channel quantum mechanics and neural networks made an appearance some time ago in connection with the problem of using adaptive optics techniques when photon noise is important. The importance of the Newton-Jost equations for this problem was first pointed out by Freeman Dyson [1975] The setup considered by Dyson is a deformable mirror surface Σ whose shape is chosen so as to just compensate for small shifts φ(σ,t) in the optical path length of light rays incident on a surface at various locations σ. Dyson proceeds by writing down equations describing the interplay between small deformations in the shape of a surface and changes in the intensity of light on an array of optical sensors which convert intensity variations into phase variations. The first equation supposes that we have a control system that adjusts the displacement δ(σ,t) of the surface with sufficient accuracy so that the intensity of light on the sensor array at time t and position x is a linear function of the error e(σ,t) ≡ δ(σ,t) + φ(σ,t)

$$I(x,t) = I_0(x) - \int d^2\sigma B(x,\sigma)e(\sigma,t) \tag{11.5}$$

where I(x,t) is the recorded intensity of light on the sensor array at time t and position x, and $I_0(x)$ is the recorded intensity on the sensor array when the surface Σ is illuminated in the absence of imposed variations in phase with respect to position or time. The second equation relates the deformation of the surface at location σ produced by the feedback control to the intensity of light on the sensor array:

$$\delta(\alpha,t) = \int d\Omega \int_{-\infty}^{t} dt' A(\sigma,x,t')I(x,t) \tag{11.6}$$

where the integral over $d\Omega$ means sampling the light intensity at a sufficiently large number of points on the sensor array as is required to determine the parameters which define the shape of the surface. When photon noise is neglected, then equations (11.5–11.6) have the classical solution (in matrix shorthand)

$$\mathbf{e} = [1 - AB]^{-1}\, \boldsymbol{\varphi} \tag{11.7}$$

Eq. 11.7 is a simple model for adaptive optics systems that use phase measurements over an aperture to adjust the position of the surface so that it locally tracks the change in optical path length. What is most remarkable though is that when photon noise is taken into account the problem of changing the shape of the surface to compensate for changes in the optical path of the illuminating beam becomes equivalent to solving the multi-channel Schrodinger equation. This situation is qualitatively different from the classical case because the negative feedback in (7) would amplify the photon noise. Dyson showed that in the presence of photon noise the optimal feedback matrices $\mathbf{A}(\sigma,x,t')$ and $\mathbf{B}(\sigma,x)$ satisfy $A = KB^T I_0^{-1}$, where $\mathbf{K}(\sigma_1,\sigma_2,t_1,t_2)$ is a causal matrix satisfying the equation

$$\mathbf{K}+\mathbf{K}^T+\mathbf{K}(\mathbf{B}^T I_0^{-1}\mathbf{B})\mathbf{K}^T+\mathbf{U}=0\,, \tag{11.8}$$

where $\mathbf{U}$ is the average $<\varphi_1\varphi_2>$ over time. As Dyson pointed out in his paper (8) is essentially equivalent to the multi-channel Newton-Jost equation. This connection between adaptive optics in the presence of photon noise and quantum inverse scattering theory is particularly interesting because it provides an explicitly quantum mechanical model for how the mammalian brain might remove the "noise" from sensory data.

One could imagine that the mammalian brain uses a neural network that effectively solves the Newton-Jost equations to extract cognitively significant information from visual images. This leads us to the question raised at the beginning. Is it conceivable that deep learning neural networks could be used to solve the many body Schrodinger equation? Of course neural networks normally involve only probabilities and not complex amplitudes. However, phase and amplitude heterodyne measurements of the state of a microwave cavity coupled to a single qubit allows one to construct the stochastic motion of a qubit on the surface of its Bloch sphere due to quantum noise [Murch *et al.*, 2013]. In much the same way that time variation in the intensity of light in Eq. 11.8 can represent the effect of photon noise, quantum walks in phase space serve as a surrogate for canonical Hamiltonian quantum dynamics. Actually it has been known for some time that phase space representations of quantum dynamics can be mapped to a set of stochastic differential equations that can be solved with Monte Carlo techniques [20]. Applications of this technique to solve many body quantum problems have been limited though because there is no known analytic technique for determining the time scale for which sampling errors lead to decreasing deviations from an exact solution. On the other hand neural networks have the advantage that one can, at least in principle, always choose the parameters of the model to provide a ML fit to observed data. Furthermore one doesn't have to have data for exactly the system that one is trying to analyze, because neural networks are capable of generalizing what they have learned to situations they have not seen before.

A very nice demonstration the how neural networks can learn to create new examples of cognitively significant curves from known examples of cognitively significant curves is provided by the application of deep learning to the recognition of hand written characters [Lake et al., 2015]. In particular, the handwriting recognition network described by Lake [2015] has the capability of generating new recognizable characters knowing only the pen strokes used to create a few examples of valid characters. One might imagine that these results could be transferred to the problem of solving the many-body Schrodinger equation by using the short-time quantum walks of particles – which can be generated with great accuracy for any system – together with the type of generative model for hand written characters described by Lake [2015] to calculate the quantum properties of a novel system. Of course this would require translating the extended quantum walks of particles into a multi-particle wave function. Fortunately this is something neural networks ought be good at doing.

As an example, if there is a molecule whose energy eigenstate wave-functions are approximately known, e.g. from density functional calculations, then one could generate short time quantum walks for all the types of particles network in the molecule – viz. the electrons in each eigenstate as well as the nuclei if one wanted to go beyond the Born-Oppenheimer approximation. One could then perhaps use a deep learning generative model similar to the model illustrated in Fig 11.4, or alternatively a neural network in which the positions of electrons and nuclei are represented by the amplitudes of quantum oscillators to simulate the real time quantum dynamics of a molecular system. Using neural network pattern recognition techniques one could then generate the energy eigenstate wave functions with correlations between all the particles taken into account.

11.4 Acknowledgements

This work was performed under the auspices of the U.S. Department of Energy by Lawrence Livermore National Laboratory under Contract DE-AC52-07NA27344.

Bibliography

Bishop, C. M. (2012). Model-based machine learning, *Phil. Trans. of The Royal Society A* 371, 2012022.

Chapline, G. (2004). Quantum mechanics and pattern recognition, *Int. J. of Quantum Information,* 02, pp. 295-303.

Chapline, G. (2001). Quantum mechanics as self-organized information fusion, *Phil. Mag. B* 81, pp. 541-549.

Dayon, P., Hinton, G.E., Neal, R., and Zemel, R. S. (1995) The Helmholtz Machine, Neural Comp. 7, pp. 1022-1037.

Dowling, M. R., et al. (2006) Monte Carlo techniques for real-time quantum dynamics, *J. Comp. Phys.* 220, pp. 549-567.

Durban, R. and Willshaw, D. (1987). An analogue approach to the traveling salesman problem using an elastic net method, *Nature,* 326, pp. 689-691.

Dyson, F. J. (1975). Photon noise and atmospheric noise in active optical systems, *J. Optical Soc. Am.* 65, pp. 551-558.

Feynman, R. P. and Hibbs, A. R. (1965) *Quantum Mechanics and Path Integrals*, Chapter 12 (McGraw-Hill, Boston).

Harris, R., et al. (2010) Experimental Investigation of an eight-qubit unit cell in a superconducting optimization processor, *Phys. Rev. B,* 82, 024511.

Hertz, J., Krogh, A. and Palmer, R. (1991) Introduction to the Theory of Neural Computation (Addison-Wesley, Redwood City CA).

Hinton, G. E., Dayan, P., Frey, B. J. and Neal, R. M. (1995) The "wake-sleep" algorithm for unsupervised neural networks, *Science* 268, pp. 1158-1161.

Jordan, M. I. and Mitchell, T. M. (2015). Machine Learning: Trends, perspectives, and prospects, *Science* 349, pp. 255-260.

Kadanoff, L. P. and Brown, A. C. (1979) Correlation Functions on the Critical Lines of the Baxter and Ashkin-Teller Models, *Ann. Phys.* 121, pp. 318-342.

Kohmoto, M. K., Dennijs, M. and Kadanoff, L. P. (1981). Hamiltonian studies of the d=2 Ashkin-Teller model, *Phys. Rev.B* 24, pp. 5229-5241.

Kohonen, T. (1995) *Self-Organizing Maps*, (Springer- Verlag, Berin).

Lake, B. M., Salakhutdinov, R. and Tenebaum, J. B. (2015). Human-level concept learning through probabilistic induction, *Science* 350, pp. 1332-1338.

Munford, D. (1994) Large-Scale Neuronal Theories of the Brain, eds. Koch C and Davis, J. L., "Neuronal Architectures for Pattern-theoretic Problems," (The MIT Press, USA) pp. 125-152.

Murch, K. W., Weber, S. J., Macklin, C. and Siddiqi, I. (2013) Observing single quantum trajectories of a superconducting qubit, *Nature,* 502, pp. 211-214 (2013).

Newton, R. G. (1989) *Inverse Schrodinger Scattering in Three Dimensions,* (Springer-Verlag, New York).

Ripley, B. D. (1996). *Pattern Recognition and Neural Networks* (Cambridge University Press, Cambridge).

Tubmanm N. M., Dubois, J. L. and Alder, B.J. (2012) Advances in Quantum Monte Carlo **1094**, eds. Tanaka, S., Rothstein, S. M. and Lester, W. A., "Recent Results in the Exact Treatment of Fermions at Zero and Finite Temperature," (American Chemical Society, USA) pp. 41-50.

Chapter 12

The Early Years of Molecular Dynamics and Computers at UCRL, LRL, LLL, and LLNL

Mary Ann Mansigh Karlsen

I'm the young woman in the picture shown in Fig. 12.1 that appeared with the invitation to the Symposium to celebrate Berni Alder's ninetieth birthday. I worked with Berni for over 25 years on the computer programs that provided the data he needed to write the fifteen papers published in scientific journals on Studies in Molecular Dynamics. My name appears at the end of each one thanking me for computer support. It has been interesting to look on the Internet to find my name in the middle of many foreign languages, including Japanese characters and Russian Cyrillic script. It shows how Berni's work has been of interest to many scientists all over the world from the earliest years. Figure 12.1 was also included with articles written when he received the National Medal of Science from President Obama in 2009.

I'll try to describe how, not only was the field of Molecular Dynamics at its very beginning, but it was also the beginning of the development of main frame computers. The Lab hired college graduates with degrees in mathematics and the sciences to become programmers. Computers were so new that the universities did not offer degrees in Computer Science until the 1960s.

The photograph shown in Fig. 12.1 was taken in the early 1960s. I am with Berni and Tom Wainwright in Berni's office located on the second floor of an old Navy barracks building. There was no air conditioning in the building. Newer buildings, including those built for the computers, had general air-conditioning but it often failed during very hot weather. Berni and most other lab scientists did not write their own computer programs. They were assigned programmers (later called computer scientists.) I started working with Berni sometime in the late 1950s. Programming at first was very specific to each new set of computers. Using new instruction sets, new codes were written in an assembly language that took advantage of the faster processors and increased memory of each new model. The computers ran 24 hours a day. Operators ran production runs at night and weekends to provide the scientists with data. The programmers would sign up for short periods of time during the day to work directly on the computers to debug their codes and set up new production runs.

Fig. 12.1. From left to right, Mary Ann Mansigh Karslen, Berni Alder, and Tom Wainwright, *c.* 1960.

In 1960 the Lab acquired the LARC (Livermore Advanced Research Computer), a new computer built to their specifications. It was the fastest computer available at the time. This was about five years after Berni had begun his Studies in Molecular Dynamics. Fig. 12.2 shows me at the LARC console during a debug session. During those years, the lab had added several new computers including IBM's 701, 704 and 709 machines in addition to the very first large main frame computer, the UNIVAC, acquired in 1952.

Fig. 12.2. Mary Ann Mansigh Karlsen sitting at the Livermore Advanced Research Computer (LARC) console.

My first assignment was with the UNIVAC group of programmers. I had answered an ad for Mathematicians listed in the MEN WANTED section of the Oakland Tribune for programmers to work at UCRL in Livermore. (The following week the ad also appeared in WOMEN WANTED.) I was the youngest and the only girl amongst 25 men, mostly World War II vets, who were hired by the Computation Department in April and May of 1955. Berni had his first computer program called APE (Approach to Equilibrium) running on the UNIVAC. I was not assigned to it; my office partner was. However, even then as now, he did not arrive at work until late morning on Tuesdays and Thursdays. When he would come to the office to check on his program, Mary Grace had often left for lunch. Since I was usually there, he would ask me to help him find his output or to make some adjustments to the program. When Mary Grace left the laboratory, it was natural for him to ask that I continue to work with the APE program.

That was the beginning of the many years of my Molecular Dynamics experience. APE, the first Molecular Dynamics program on the UNIVAC, was limited to a small number of particles. It had been programmed by Shirley Campbell who had since left the Lab. I soon became familiar with its details and would add additional calculations to it. Later I took over the Molecular Dynamics program called STEP, which was coded by Norm Hardy to run on the first IBM machines. They had more memory and were faster than the UNIVAC so Berni was able to get additional data using a larger number of particles.

Fig. 12.1 shows much of what computation was like in 1960. The big stacks of paper illustrate the fact that for many years that was the only means of obtaining large amounts of data from the computers. Berni and his colleagues had to use the data printed out on paper to analyze the results they were looking for. Punched cards (not shown) were used to enter new programs into the computers. The computer then read and wrote on tapes, like the ones sitting on his desk, to store the code and restart data. The output data from the computer was written on other tapes for printing. Over the years many different larger and faster printers were developed that could quickly print the vast amounts of data generated each day. The first printer for the UNIVAC output was a modified teletypewriter equipped with a printer board to format the data.

The LARC came with an oscilloscope and a Polaroid camera so that data could be recorded on film and looked at during a run. Berni made use of this feature to actually see how the particles moved by recording their positions over a period of time. In Fig. 12.1 we are looking at some of these computer generated pictures. Fig. 12.3 is an example of one of the photographs we made with this process. This image of a two dimensional system shows the movement of the center of the particles and how starting from a regular pattern they slowly begin to melt, going from a solid state to a liquid. From the various prints on the desk we could see how larger particles remain in their structured solid positions, smaller particles move slowly over time and leave their original locations as they go to a liquid state, and the smallest particles representing a gas move around much more freely.

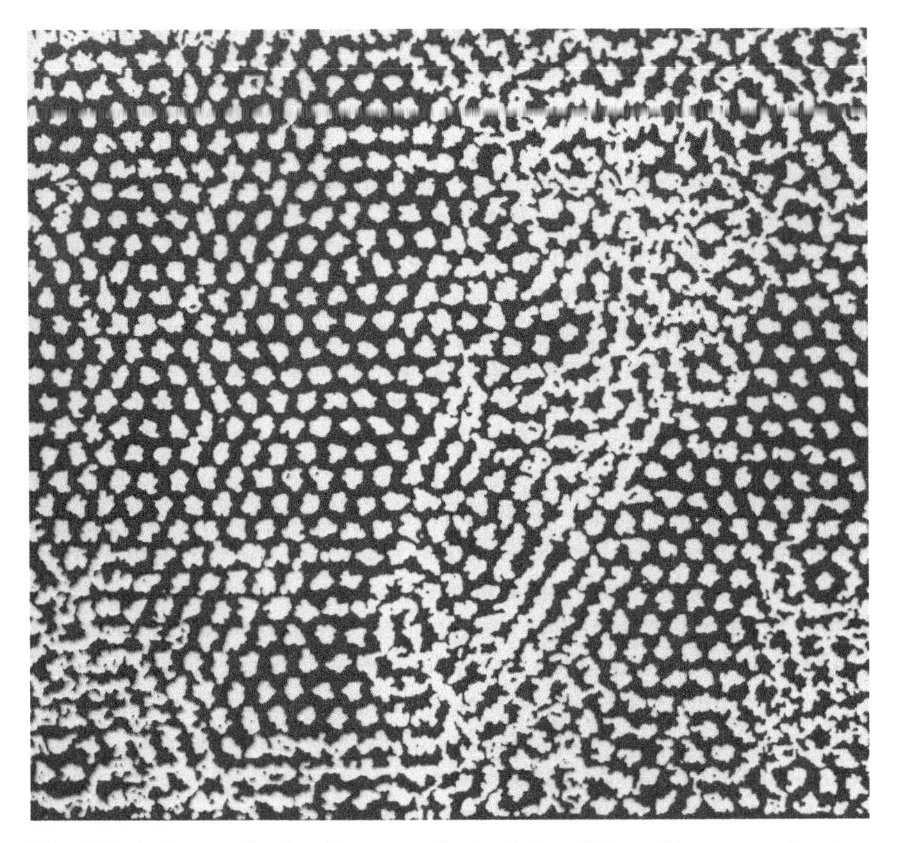

Fig. 12.3. A photograph taken from an early simulation of the melting process in a two dimensional system of particles.

Berni, always ready to use new features, suggested making a movie on the LARC. He wanted to illustrate irreversibility by recording the movement of a few particles over a period of time. The particles were arranged in an orderly pattern on one side of an empty box. Using the STEP program, we recorded the movement of the particles as they filled the entire space. We then stopped the action and reversed all the velocities. The particles returned to their initial positions. I had no precedent to follow, but with the instructions that the frames in a movie should be 16 seconds apart, using the data generated on the computer, I proceeded to represent each particle as a circle on each frame. It took many tries to record each particle, one dot of the circumference at a time, and to follow them as they collided with each other or the side of the box without distorting their shape. We were successful and Berni provided a loop film that accompanied a physics textbook used at UC Berkeley.

Not being a physicist, I thought of the STEP program in rather simple terms. It basically always started by arranging the hard spheres in a fixed pattern in a one by one by one box in three dimensions, giving the particles random velocities in each direction and "following the balls as they bounced around in the box." The program kept track of the collisions that occurred and the movement of the particles over time. We also did this in two dimensions. This pattern continued in all our calculations throughout the years. The number of hard spheres varied, the size of the particles varied, and other factors were calculated for additional information as Berni and others added more details to their molecular studies.

Since my bachelor's degree was in mathematics and chemistry and I had only the basics of classical physics, my interests were more computer-oriented and not in the analysis of the results. The early computers each had their own set of instructions, so the same program was completely re-coded to take advantage of any new features. When FORTRAN became available, programming changed. The same written codes could be used for all the computers. My programming emphasis then concentrated on how to make the programs very efficient. Another big change in using the computers came when several programs could run on the computer at the same time. Depending on their priority, the highest priority would run until it had a waiting period, when the next priority would run until the higher priority was ready. Because the STEP program was well established and did not use much space in memory, it was available at the lowest priority to fill any unused time. Berni had many variations of his Molecular Dynamics studies to allow STEP to run on all the computers. The STEP program often was one of the last to run on older slower computers and also one of the first to run on any new computer.

The Lab acquired many of the newest and fastest computers as they became available and often kept them for many years. Beside the Univac and LARC, IBM provided most of the early computers starting with the first two 701s. This went on to include four 704s, four 709s, four 7090s, and five 7094s, before the first CDC 1604 computer arrived from Control Data Corporation in 1962. The Lab later acquired another 1604, two CDC 3600s and four 6600s. That was a lot of computers, but since Berni's policy was to use up to 40 hours of computer time for each run, he wanted to use as much idle time as possible. Because of this, Berni's STEP program was the largest user of computer time at Livermore for many years. But that wasn't quite enough for the many problems he was solving. When the laboratory in Berkeley, LBNL, acquired their first CDC 6600s, Berni realized there would be a lot of idle time before they were completely utilized. He requested and was given permission to run his STEP jobs on their machines. Since I lived in Berkeley, it was easy for me to make trips up the hill to set up runs and collect output during that time.

In the early 1970s a system called OCTOPUS was developed using cables to connect the network of computers with the many users. Programmers could remain in their offices while working on the computers. Using teletypes and CRT monitors we could write our programs, debug them on the computers, and follow the progress of production runs from our offices. The teletypes and CRTs were later replaced by personal computers when they became available. We no longer spent time in the computer room. We moved into offices scattered around the Lab. The entire programming environment had changed.

By the 1980s the new group of scientists working with Berni could write their own programs and control their work directly on the newest and fastest computers. The fields of studies also had advanced and become more complicated. Berni's group began using quantum mechanics and the Monte Carlo method. I did not have an advanced physics background, so my many years of working with Berni came to an end. My programming career in Molecular Dynamics was over and I went on to work a few more years with other Lab scientists before retiring.

I was pleased to have been included among the special guests at Berni's 90th Birthday Symposium. I enjoyed the years I worked with him and have found much satisfaction knowing my computer work contributed to his success in the field of Molecular Dynamics.

Chapter 13

Overcoming the Fermion Sign Problem in Homogeneous Systems

Jonathan L DuBois[a], Ethan W. Brown[b], and Berni J. Alder[a]
[a]*Lawrence Livermore National Laboratory, Livermore, CA 94550, USA*
[b]*Department of Physics, University of Illinois at Urbana-Champaign, 1110 W. Green St., Urbana, IL 61801-3080, USA*

Abstract

Explicit treatment of many-body Fermi statistics in path integral Monte Carlo (PIMC) results in exponentially scaling computational cost due to the near cancellation of contributions to observables from even and odd permutations. Through direct analysis of exchange statistics we find that individual exchange probabilities in homogeneous systems are, except for finite size effects, independent of the configuration of other permutations present. For two representative systems, ^3He and the homogeneous electron gas, we show that this allows the entire antisymmetrized density matrix to be generated from a simple model depending on only a few parameters obtainable directly from a standard PIMC simulation. The result is a polynomial scaling algorithm and up to a 10 order of magnitude increase in efficiency in measuring fermionic observables for the systems considered.

Path integral Monte Carlo (PIMC) methods provide essentially exact results for low temperature properties of N-body Bosonic systems.[1] While the same algorithm can be applied to Fermions, a sign problem arising from the approximately equal weights of the $N!$ oppositely signed permutations, limits the accuracy of the results. In fact, naive application of the PIMC method to fermions results in exponentially decreasing efficiency as the temperature decreases and N increases.[2] Consequently, enforcement of

Fermi symmetry for all but the smallest finite temperature systems has so far required the introduction of an approximation that restricts path integrals to prevent sign changes, analogous to the fixed node approximation used in ground state quantum Monte Carlo (QMC).[3]

In order to determine whether it is possible to overcome the sign problem directly, we have used the PIMC method to examine permutation space with great accuracy. From this data we are able to show that for homogeneous systems, the effective dimensionality of the sum over permutations can be reduced to a relatively small finite number, allowing for exact treatment of large systems to low temperatures.

The diagonal density matrix of a fermionic system at temperature $T = 1/k\beta$, denoted $\rho^{\mathcal{A}}(\mathbf{R};\beta)$, can be expressed as a weighted sum over off-diagonal distinguishable density matrices, ρ^D, connecting the many-body coordinates, $\mathbf{R}$, to all permutations $\mathcal{P}(\mathbf{R})$ in imaginary time β. This gives

$$\rho^{\mathcal{A}}(\mathbf{R};\beta) = \frac{1}{N!} \sum_{\mathcal{P}\in S_N} (-1)^{\mathcal{P}} \rho^D(\mathbf{R}, \mathcal{P}(\mathbf{R});\beta). \tag{13.1}$$

where S_N is the symmetric group of all possible permutations of N particles.

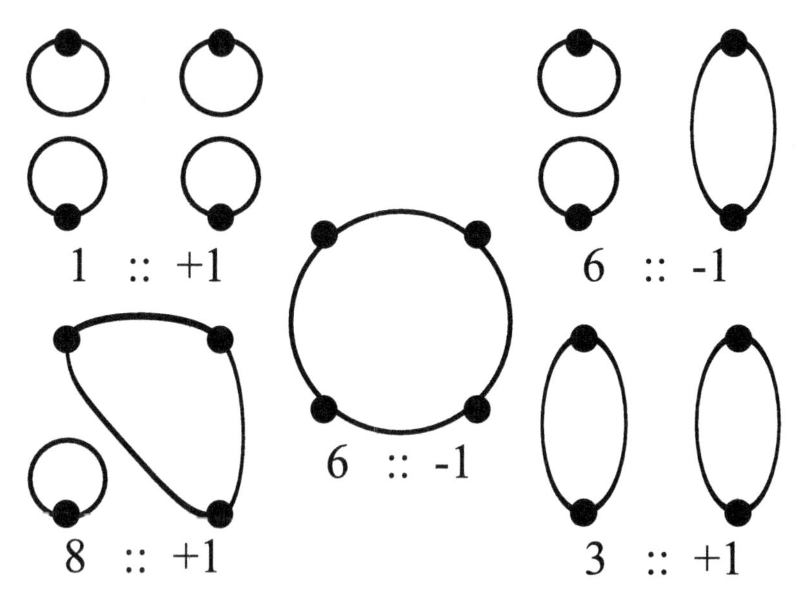

Fig. 13.1 Diagramatic representation of the equivalence classes of the symmetric group for 4 particles, S_4, the number of elements in each class and the sign of the contribution of members of each class to the antisymmetric partition function are shown below each diagram.

Direct treatment of this sum over permutations is made difficult in two ways: the high dimensionality of the full quantum many-body density matrix necessitates the use of a stochastic sampling method for evaluation of expectation values of observables, and the alternating sign arising from antisymmetrization over permutations of paths $(-1)^{\mathcal{P}}$ results in a large variance. Our goal in this work is to reduce the effective dimensionality of the sum over permutations. As a starting point along these lines we note that evaluation of (13.1) can be simplified by recognizing that while there are $N!$ possible permutations of N particles, the symmetry group S_N can be further organized into subsets of topologically equivalent diagrams $[p] \equiv \{\mathcal{P} \in S_N | \mathcal{P} \sim p\}$.[4] Figure 13.1 shows representative members from the five equivalence classes of the symmetric group for four particles, S_4.

Expectations evaluated within different members of the same class are identical, and one need only evaluate a single representative permutation in each class rather than the full sum. The observables for each equivalence class $[p]$ may be written as

$$\langle \mathcal{O} \rangle_{[p]} = \frac{\sum_{\mathcal{P}\in[p]} \int d\mathbf{R}\ \hat{\mathcal{O}}\rho(\mathbf{R}, \mathcal{P}(\mathbf{R}); \beta)}{\sum_{\mathcal{P}\in[p]} \int d\mathbf{R}\ \rho(\mathbf{R}, \mathcal{P}(\mathbf{R}); \beta)} \tag{13.2}$$

and may be measured independently during simulation. To then reconstruct the fully antisymmetrized observable one must sum over the equivalence classes with their respective sign,

$$\langle \mathcal{O} \rangle = \frac{\sum_{[p]} \sigma_{[p]} Z_{[p]}(\beta) \langle \mathcal{O} \rangle_{[p]}}{\sum_{[p]} \sigma_{[p]} Z_{[p]}(\beta)} \tag{13.3}$$

where we have defined $Z_{[p]}(\beta)$ as the contribution of a given equivalence class to the full symmetric partition function and $\sigma_{[p]}$ is the sign associated with the class. At zero temperature, the denominator of (13.3) becomes exactly zero. At any finite temperature, each nonequivalent permutation sector will have a different mean energy and therefore a different probability and a nonzero contribution to the antisymmetrized partition function.

The cost of summing over all unique equivalence classes of S_N still scales exponentially with N and attempts to make use of the structure afforded by the symmetry group directly have met with mixed success.[5,6] Ultimately, in order to obtain a polynomial scaling algorithm, it is therefore essential to determine whether all sectors need to be evaluated equally or at all. To this end, we outline a scheme to dramatically reduce the effective dimensionality of the sum over permutations.

Each permutation class is uniquely identified by the number of loops of a given length so that $[C_1, C_2, \ldots, C_N]$ represents the class with C_1 cycles of length one, C_2 cycles of length two, and so on. In a noninteracting system all sectors can be further factored into products over cycle lengths, ℓ, allowing the contributions to the total partition function from each sector to be written as,[7]

$$Z_{[p]}(\beta) = \prod_{\ell=1}^{N} M_{[p]\ell} P_\ell(\beta)^{C_{[p]\ell}} \tag{13.4}$$

where $M_{[p]\ell} \equiv 1/(C_{[p]\ell}!\ell^{C_{[p]\ell}})$ is a combinatorial factor, and P_ℓ is related to the single-particle partition function and is independent of the equivalence class $[p]$. Given this expression for $Z_{[p]}$, observables take a particularly simple form. For example, taking the β derivative of the partition function, one finds[7]

$$\langle E(\beta)\rangle_{[p]} = \sum_{\ell=1}^{N} C_{[p]l} E_\ell(\beta) \tag{13.5}$$

where E_ℓ is the contribution to the toal energy from particles participating in a cycle of length ℓ. The total energy $\langle E\rangle$ is then given by (13.3). A similar expression can be constructed for the pair correlation function and other observables.[7]

The significance of (13.4) and (13.5) within the context of the sign problem in PIMC is that expectation values can be obtained by evaluating the N positive definite expectations $P_\ell(\beta)$ rather than the $N!$ terms in (13.1). The net result is an algorithm with an effective computational cost scaling as well as $\mathcal{O}(N^2)$ in the number of particles since the probability density associated with each of the equivalence classes can be reconstructed from the probability densities of an $\mathcal{O}(N)$ subset of S_N (e.g. $[N, 0, 0, \ldots], [0, N/2, 0, 0, \ldots], [0, 0, N/3, 0, \ldots], etc.$) and the computational cost of the PIMC algorithm itself can be $\mathcal{O}(N)$.

Going further, it can be shown that for a noninteracting (ideal) gas the contribution to the partition function of neighboring sectors is proportional to their length[7] allowing one to relate P_ℓ to an exponentially decreasing function in cycle length so that $P_\ell = p_2^{-\ell}$. This single expectation, p_2, can be seen as the mean probability of a permutation between two particles. Taking advantage of these observations allows for a reduction of the task of finding the relative probability of different sectors to that of determining a single temperature dependent value and an $\mathcal{O}(N)$ algorithm.

In an interacting system one might assume that the relative probability of different cycle lengths P_ℓ will depend on the permutation sector $[p]$ and construct a sector dependent $P_{[p]\ell}$ as an expansion around an averaged $\bar{P}_\ell$ such that

$$P_{[p]\ell}/\bar{P}_\ell = \quad 1 + \sum_{m=1}^{N} C_{[p]m}(\tfrac{\bar{P}_{\ell m}}{\bar{P}_\ell} - 1) + \sum_{m,n=1}^{N} C_{[p]m}C_{[p]n}(\tfrac{\bar{P}_{\ell mn}}{\bar{P}_\ell} - 1) + \dots \tag{13.6}$$

where $\bar{P}_{\ell m \dots}$ is related to the probability of finding a cycle length ℓ given the existence of cycles of lengths $m \dots$. Krauth and Holzmann have shown that for weakly interacting bosons effective interactions between different cycle lengths, i.e. $\bar{P}_{\ell m \dots}$, are negligible[8,9] suggesting that determination of $\bar{P}_\ell$ alone is sufficient for a wide class of systems *. In this work we have found through direct PIMC simulation that this same simple structure is obeyed in two important strongly interacting fermi systems. Our key finding is that, aside from small finite size effects, the contribution of a loop of length ℓ to the probability of a permutation sector is the same for all permutation sectors. This remarkable result leads to a dramatic reduction in the computational cost required to evaluate any signed observable.

In what follows, we present results of this approach applied to two prototypical strongly interacting fermonic liquids – the homogeneous electron gas (HEG) and liquid ^{3}He. The character of exchange interactions in these two systems represent two qualitative extremes. In the HEG, a weak correlation hole results in a high probability of exchange between nearest neighbors. In contrast, ^{3}He has a strong correlation hole resulting in a significantly lower nearest neighbor exchange probability. As a consequence of its higher compressibility and higher exchange probability, we find that the permutation structure of the HEG more closely resembles the noninteracting gas and the average value of the sign in the HEG, $\langle\sigma\rangle$, decays to zero more rapidly with decreasing temperature and increasing number of particles than in ^{3}He. The strong correlation hole in ^{3}He has the effect of modifying the combinatorial factor, $M_{[p]\ell}$, in (13.4) for a finite simulation box requiring the addition of a model to account for the absence of overlapping exchange loops.

We have examined both a low density ($r_s = 10$) and high density ($r_s = 1$) state of the HEG both at the Fermi temperature, T_F, and at 1/8

*The idea that the qualitative features of the loop structure may be obtained from the pair exchange probability alone was already proposed by Feynman[10] to describe the λ transition in liquid ^{4}He. In order to evaluate the weight of each permutation sector analytically, Feynman assumed that the number of ways of forming a closed loop consisting of k particles is approximately independent of the arrangement of other loops and could be accounted for by a single effective parameter. This results in the relative probability of different permutation sectors being determined by a simple Poisson distribution.

Table 13.1 Total energies per particle for 33 spin-polarized electrons at $r_s = 1, 10$ and $T/T_F = 0.125, 1.0$. From left to right, we plot energy estimates for standard signful PIMC, restricted PIMC from,[11] reconstructed PIMC using P_ℓ as in (13.4), and reconstructed PIMC using p_2.

r_s	T/T_F	$PIMC$	$RPIMC$	P_l	p_2
1.0	0.125	1(10)	2.35(1)	2.3(1)	2.33(6)
1.0	1.0	3(7)	8.69(3)	8.7815(7)	8.7801(7)
10.0	0.125	−0.0(1)	−0.1038(2)	−0.1030(1)	−0.1033(1)
10.0	1.0	−0.040(2)	−0.0403(5)	−0.0402(1)	−0.04025(5)

T_F. Table (13.1) summarizes energies of the HEG for these two densities and temperatures compared with previous exact and fixed-node results.[11] We find that our reconstructed energies, both by fitting p_2 and P_ℓ directly, match well with previous fixed-node results. This agreement provides a direct demonstration that the permutation structure of the HEG is well described by free fermions with an effective mass. Results obtained by fitting P_ℓ are within error bars of those obtained with the more constrained p_2 fits. Previous exact simulations were not possible below the Fermi temperature, giving estimates for the energy with a variance larger than the value itself, while both reconstructions work well at $1/8\ T_F$. Our new method thus extends the regime where unbiased exact simulations are possible to much lower temperatures. The increase in efficiency is most notable for $rs = 1.0$ and $T/T_F = 1.0$ where resummation of the same PIMC data using the p_2 model results in a standard error 5 orders of magnitude smaller than the direct antisymmetrized sum. Given that the statistical error scales with the square root of the number of independent samples, an additional 10^{10} times as many samples would be required to obtain the same result by direct summation. For some of the points examined, we see up to $\sim 1\%$ discrepancies with the fixed-node result. It is tempting to assume that the current results are more accurate since they avoid the fixed node approximation. However, in order to confirm this one would need to include higher order terms in the cluster expansion (13.6) and demonstrate convergence.

Results of our approach applied to liquid ^{3}He are shown in Figure 13.2. An approximate (truncated) direct summation over permutations performed in a previous work[12] (solid circles) agrees well with experimental data[13] (dashed line) down to temperatures well below the ^{3}He Fermi temperature, $T_f = 1.7K$. Reconstructed energies were obtained with an ≈ 10 order of magnitude reduction in computational cost compared to a naive direct summation over partitions and agree to within statistical errorbars with the experimental values as opposed to the approximation introduced by the restricted path method.[3]

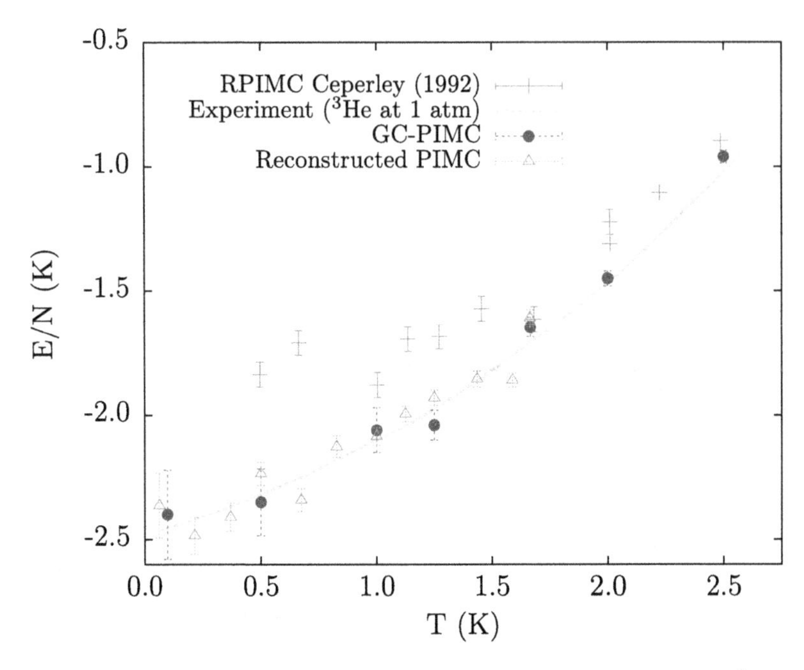

Fig. 13.2 Results of antisymmetrized grand canonical PIMC applied to liquid ^{3}He. Our direct results (solid circles) and the reconstructed energies described in the text (open triangles) agree well with experimental data (dashed line) down to temperatures well below the ^{3}He Fermi temperature. Results obtained with restricted PIMC in[3] (+ signs) are shown for comparison.

The strong correlation hole in ^{3}He results in a deviation of the combinatorial factor $M_{[p]\ell}$ in (13.4) in finite systems from the free gas (overlapping permutation loops) model. In order to account for this and extract a thermodynamic limit value for p_2 from our finite-N PIMC data, we numerically solve an analogous but discrete problem namely, a nearest neighbor Ising model with N sites and a nearest neighbor connectivity close to that of the liquid for each distinct permutation sector.†

Figure 13.3 shows the values of p_2 for a range of permutation sectors for $N = 66$ ^{3}He at $T = 1.2$ Kelvin. Solid squares in the figure show results for p_2 obtained based on an uncorrected Poisson model. The statistically

†We note that the choice for the Ising lattice is somewhat ad hoc in the sense that if a square lattice is assumed, where each He atom has 6 nearest neighbors whereas an e.g. hcp lattice would have 12 neighbors. However, we have found that while the choice of lattice does change the mean extracted value of p_2 somewhat, the reconstructed values of $Z_{\mathcal{P}}$ do not depend significantly on the choice of lattice, being insensitive to the detailed form of the lattice for the temperatures we have considered.

significant drift in the mean value of p_2 with increasingly long permutation cycles resulting from finite size effects is evident. In contrast, results for p_2 obtained from the inverse Ising model are found to agree with the mean value obtained over all sectors within statistical error bars ($\approx 1\%$). The mean value of p_2 obtained in this way was used to weight expectation values over the full antisymmetrized density matrix.

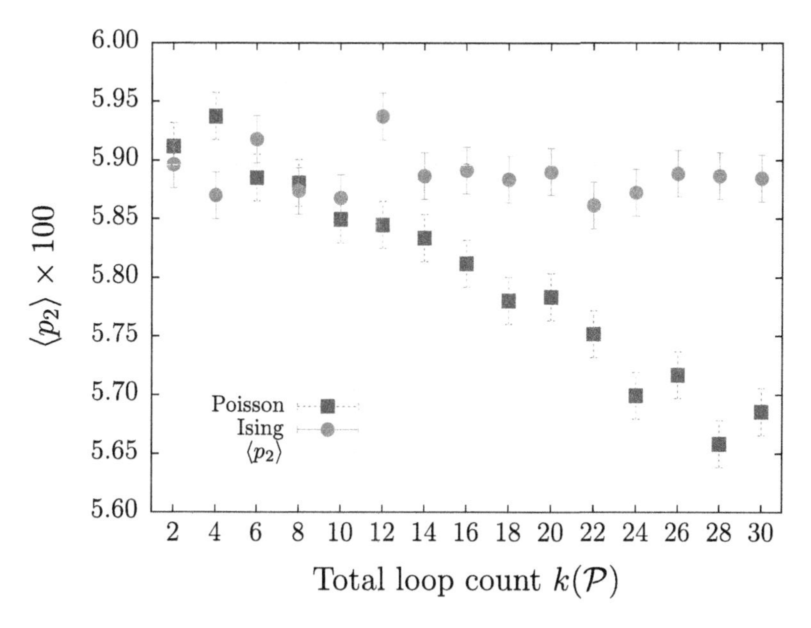

Fig. 13.3 Representative values of the extracted pair exchange probability p_2 across a representative subset of equivalence classes containing increasingly long loops using Poisson statistics (squares) and the numerically inverted Ising model (circles). The dashed line shows the mean value $\langle p_2 \rangle$ over all sectors obtained via the Ising method.

In conclusion, we have shown that it is possible to directly address the sign problem for homogeneous systems by taking advantage of their relatively simple permutation space structure. While here we report only energies, we note that other diagonal observables may be accessed in the same way.[7] Current efforts are focused on reconstructing the permutation space of heterogeneous systems in the hope that we can also exactly calculate properties of general many-body Fermion systems much below the Fermi temperature.

Acknowledgements

This work was performed under the auspices of the U.S. Department of Energy by Lawrence Livermore National Laboratory under Contract No. DE-AC52-07NA27344 and supported by LDRD Grant No. 10-ERD-058 and the Lawrence Scholar program.

References

1. D. M. Ceperley, Path integrals in the theory of condensed helium, *Rev. Mod. Phys.* **67**, 279 (1995).
2. D. M. Ceperley, *Path integral Monte Carlo methods for fermions*, in *Monte Carlo and Molecular Dynamics of Condensed Matter Systems*, eds. K. Binder and G. Ciccotti (Editrice Compositori, Bologna, Italy, 1996), Bologna, Italy.
3. D. M. Ceperley, Path-integral calculations of normal liquid ^{3}He, *Phys. Rev. Lett.* **69**, 331 (1992).
4. R. Feynman, *Statistical Mechanics* (Benjamin, 1972).
5. A. P. Lyubartsev and P. N. Vorontsov-Velyaminov, Path-integral monte carlo method in quantum statistics for a system of N identical fermions, *Phys. Rev. A* **48**, 4075 (1993).
6. M. A. Voznesenskiy, P. N. Vorontsov-Velyaminov and A. P. Lyubartsev, Path-integral expanded-ensemble monte carlo method in treatment of the sign problem for fermions, *Phys. Rev. E* **80**, p. 066702 (Dec 2009).
7. J. L. Dubois, E. W. Brown and B. Alder, Supplemental material (2014), arXiv:1409.3262.
8. M. Holzmann and W. Krauth, Transition temperature of the homogeneous, weakly interacting bose gas, *Phys. Rev. Lett.* **83**, 2687 (1999).
9. W. Krauth, *Stat. Mech.: Alg. and Comp.* (Oxford University Press, 2006).
10. R. P. Feynman, Atomic theory of the λ transition in helium, *Phys. Rev.* **91**, 1291 (1953).
11. E. W. Brown, B. K. Clark, J. L. DuBois and D. M. Ceperley, Path-integral monte carlo simulation of the warm dense homogeneous electron gas, *Phys. Rev. Lett.* **110**, p. 146405 (Apr 2013).
12. N. M. Tubman, J. L. DuBois and B. J. Alder, *Recent results in the exact treatment of fermions at zero and finite temperature*, in *Advances in Quantum Monte Carlo*, eds. S. Tanaka, S. M. Rothstein and W. A. Lester (ACS, 2012), ch. 1, pp. 41–50.
13. D. S. Greywall, Specific heat of normal liquid ^{3}He, *Phys. Rev. B* **27**, 2747 (1983).

Printed in the United States
By Bookmasters